They Came

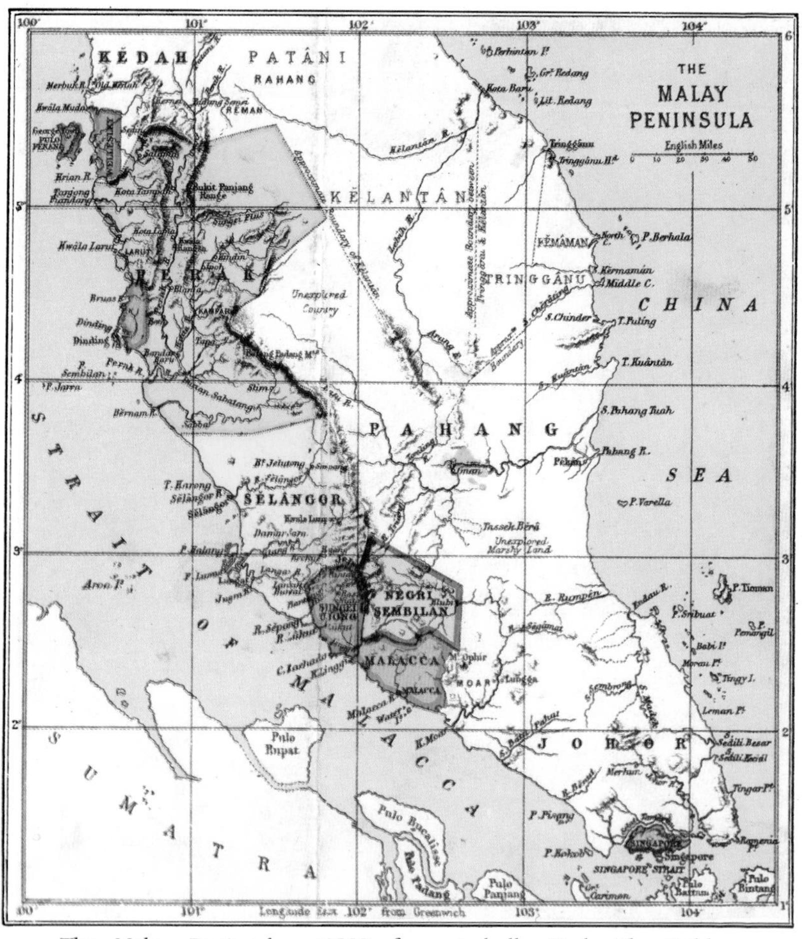

The Malay Peninsula, *c.*1880, from Isabella Bird, *The Golden Chersonese and the Way Thither*, John Murray, London, 1883.

They Came to Malaya
A Travellers' Anthology

Compiled and Introduced by

J. M. GULLICK

SINGAPORE
OXFORD UNIVERSITY PRESS
OXFORD NEW YORK
1993

Oxford University Press, Walton Street, Oxford OX2 6DP
Oxford New York Toronto
Delhi Bombay Calcutta Madras Karachi
Kuala Lumpur Singapore Hong Kong Tokyo
Nairobi Dar es Salaam Cape Town
Melbourne Auckland Madrid

and associated companies in
Berlin Ibadan

Oxford is a trade mark of Oxford University Press

Published in the United States
by Oxford University Press, New York

ISBN 0 19 588604 6

British Library Cataloguing in Publication Data

Data available

Library of Congress Cataloging-in-Publication Data

They came to Malaya: a travellers' anthology/compiled and introduced by J. M. Gullick.
p. cm.
ISBN 0-19-588604-6 (paper):
1. Malaya—Description and travel. 2. Travellers—Malaysia—Malaya—Biography. I. Gullick, J. M.
DS592.4.T54 1993
915.95'1043—dc20
92-45022
CIP

Typeset by Indah Photosetting Sdn. Bhd., Malaysia
Printed by Peter Chong Printers Sdn. Bhd., Malaysia
Published by Oxford University Press Pte. Ltd.,
Unit 221, Ubi Avenue 4, Singapore 1440

Acknowledgements

THE source of every passage is given at the end of the passage.

The Author and the Publishers gratefully acknowledge the permission to reproduce passages (identified below by their serial numbers in the anthology) given by the following copyright holders:

Campbell, Thomson & McLaughlin Ltd. (92*)
W. & R. Chambers Publishers Ltd. (99)
Chapman & Hall Ltd. (31)
Mr Choo Weng-Choon (20)
The C. W. Daniel Company Ltd. (97)
Faber and Faber Ltd. and the Standard Chartered Bank (28)
HarperCollins Publishers Ltd. (29,* 53,* 56, and 79*)
William Heinemann Ltd. (per R. I. B. Library, Reed Book Services) (55)
Miss A. L. Howe and Mr E. Howe (93)
Kelab Golf Diraja Selangor (80*)
Dr G. E. D. Lewis (54)
Mr D. Moore (33)
Malaysian Branch of the Royal Asiatic Society (23, 37, 70, 77,* and 78)
Penerbit Fajar Bakti Sdn. Bhd. (54)
The New Straits Times Press (Malaysia) Berhad (80* and 83*)
Tan Sri Dato' Dr Haji Mubin Sheppard (77*)
Mrs K. Sim (48)
Mr T. C. Skeat (5, 40, and 41)
Mrs D. Sparkes (100)
Yale University Press (61)

(* part of the passage only)

The Author has made every effort to discover which passages included in this anthology are subject to copyright, and in those cases to trace and secure the permission of the copyright holders for their inclusion. In a few cases these efforts have not been successful, and the Author and the Publishers offer their apologies to the copyright holders (if any), trusting that they will accept the will for the deed.

Contents

Wars and Troubled Times

Introduction

J. M. GULLICK

MALAYA—the peninsular part of Malaysia—is a very interesting country, which has centuries of recorded history and fascinating and contrasting traditional lifestyles among its people. This anthology has been compiled as a kind of verbal snapshot album which provides some glimpses of Malaya in recent times when the modern community was taking shape. It is a retrospective view of a century or more of Malaya as it was not long ago. Its purpose is to illustrate the past and to entertain the reader, not to give a history lesson.

In calling the collection a *Travellers' Anthology* we, the publishers and the editor, had two aspects in view. Since long-distance travel began, Malaya has stood at the cross-roads of South-East Asian communications. When Marco Polo returned from China at the end of the thirteenth century he came by sea, passing through the Strait of Malacca—though he says tantalizingly little about it. A hundred years ago or so, the celebrated Victorian lady traveller, Isabella Bird, came to Singapore on her return from Japan to Europe, and she accepted with alacrity an invitation to break her journey and visit the Malay States, then an almost unknown area even in the Straits Settlements towns. She at least was prompted to write a book, from which Passages 1, 8, and 36 have been included in this anthology, giving one of the best accounts we now have of central Malaya in 1879.

But the selection has not been limited to passages from the writings of travellers from afar. The first group of passages illustrates how arduous and varied was travel within the country, which at that time did not possess its present excellent communications. So in this anthology a 'traveller' includes anyone on the move who has some novel and vivid impression or experience to offer of Malayan travel, together with the impressions of those who crossed the paths of others on the move. One such case (Passage 62) is the picture of Stamford Raffles, one of the most visionary of Englishmen in the East, staying at Malacca before moving on to Java in 1811. By chance he took into his service a Malay youth of fifteen, who was destined to become one of the most celebrated of Malay authors, Munshi Abdullah.

In another sense this is a 'travellers' anthology' because it is intended to offer to the modern traveller, visiting Malaya, some pictures of its recent past to add to his or her recent impressions.

No anthology can include everything which the compiler would like to put into it. But, within the limits of reasonable size, this is a wide-ranging selection. Some passages are directly or in translation from the writing of Malayans. Almost all of them are about Malayans, their customs, or the places they lived in or the lifestyle of their times. Here are Malay Rulers organizing splendid entertainments for their visitors (Passage 73), and—at the other extreme—a village headman (Passage 45) giving shelter on his veranda, under its leaky roof, to the American zoologist, William Hornaday, who had come in search of specimens. There are passages about people, places, centres of worship, hospitals, schools, plantations, and wild animals. If there are passages about curious individuals and events, there are also accounts of ordinary experience, such as moving into a new house, playing games, and finding amusements and recreation.

As this is not a systematic primer but—it is hoped—a source of entertainment, the selected passages, grouped broadly by theme, each stand on their own merits. One may lead on to the next, as in the sequence on Kuala Lumpur (Passages 21–29), but more often what follows is intended to be a contrast with its predecessor. As most of the selection is about things which have changed, sometimes very rapidly, over the years, there is a short introductory note to each passage to put it in its context and explain allusions which may otherwise be a little baffling.

To preserve the authenticity of passages which were written, in many cases in the last century, the original spelling, punctuation, etc. has been retained. But, where some brief explanation of Malay words or technical terms seemed necessary, a gloss (in brackets) or a footnote has been added. To keep the length of passages within bounds, some detailed or less interesting material, footnotes and so on in the original has in places been omitted—and this is indicated by '...'. The drawings have been chosen to illustrate the passages among which they appear. They are not in every case from the books from which the passages themselves are taken, but they are contemporary with them.

After each passage comes a citation of its source, partly so that the reader may pursue his reading if he wishes to do so. Elsewhere formal acknowledgement is made to the author and publisher, or to some other person if enquiries suggest that he owns or controls the copyright. This quest is not easy, owing to the manifold changes in the modern publishing world, but every reasonable effort has been made to deal adequately with this aspect of the collection.

If it has given pleasure to the reader, or added to his interest or even provided information which he wants to have, the anthology has justified itself by enlarging the reader's appreciation of Malaya and its people, and how it and they come to be what they are. The compilation of an anthology of unfamiliar descriptions requires a wide coverage of books and other sources.The compiler gratefully acknowledges his debt to the publishers and to Ray Parker, a collector of Malayan books, for many helpful suggestions on the text and the illustrations.

Through the Jungle

1
A Ride on an Elephant

ISABELLA BIRD

Isabella Bird (1831–1904) was a celebrated traveller and author of books about her visits to remote places. At the beginning of 1879 she reached Singapore, returning to Europe from northern Japan. Here she accepted an invitation to interrupt her journey by making a short tour of Sungei Ujong (part of Negri Sembilan), Selangor, and Perak, the Malay States which had come under British control in 1874. After a brief stay at Taiping she moved on, over the Bukit Gantang pass, to Kuala Kangsar, to be the guest of Hugh Low, the British Resident of Perak. As there were no roads and no vehicles, an overland journey had either to be made on foot or by riding on the back of an elephant (or less frequently being carried in a palanquin). In the episode here described Isabella Bird abandoned an uncomfortable ride on a half-trained elephant to complete the final eight miles on foot.

I waited for the elephant in a rambling empty house, and Malays brought pierced coco-nuts, buffalo milk, and a great *bouquet* of lotus blossoms and seed-vessels, out of which they took the seeds, and presented them on the grand lotus leaf itself. Each seed is in appearance and taste like a hazel-nut, but in the centre, in an oval slit, the future lotus plant is folded up, the one vivid green seed leaf being folded over a shoot, and this is intensely bitter.

The elephant at last came up and was brought below the porch. They are truly hideous beasts, with their gray, wrinkled, hairless hides, the huge ragged 'flappers' which cover their ears, and with which they fan themselves ceaselessly, the small mean eyes, the hideous proboscis which coils itself snakishly round everything; the formless legs, so like trunks of trees; the piggish back, with the steep slope down to the mean, bare tail, and the general unlikeness to all familiar and friendly beasts....

Before I came I dreamt of howdahs and cloth of gold trappings, but my elephant had neither. In fact there was nothing grand about him but his ugliness. His back was covered with a piece of raw hide, over which were several mats, and on either side of the ridgy backbone a shallow basket, filled with fresh leaves and twigs, and held in place by ropes of rattan. I dropped into one of these baskets from the porch, a young Malay lad into the other, and my bag was tied on behind with rattan. A noose of the same with a stirrup served for the driver to mount. He was a Malay, wearing only a handkerchief and *sarong*, a gossiping, careless fellow, who jumped off whenever he had a chance of a talk, and left us to ourselves. He drove with a stick with a curved spike at the end of it, which, when the elephant was bad, was hooked into the membranous 'flapper', always evoking the uprearing and brandishing of the proboscis, and a sound of ungentle expostulation, which could be heard a mile off. He sat on the head of the beast, sometimes cross-legged, and sometimes with his legs behind the huge ear covers. Mr Maxwell assured me that he would not send me into a region without a European unless it were perfectly safe, which I fully believed, any doubts as to my safety, if I had any, being closely connected with my steed.

The mode of riding was not comfortable. One sits facing forwards with the feet dangling over the edge of the basket. This edge soons produces a sharp ache or cramp, and when one tries to get relief by leaning back on anything, the awkward rolling motion is so painful, that one reverts to the former position till it again becomes intolerable. Then the elephant had not been loaded 'with brains', and his pack was as troublesome as the straw shoes of the Japanese horses. It was always slipping forwards or backwards, and as I was heavier than the Malay lad, I was always slipping down and

Isabella Bird on her elephant, from Isabella L. Bird, *The Golden Chersonese and the Way Thither*, John Murray, London, 1883.

trying to wriggle myself up on the great ridge which was the creature's backbone, and always failing, and the mahout was always stopping and pulling the rattan ropes which bound the whole arrangement together, but never succeeding in improving it.

Before we had travelled two hours, the great bulk of the elephant without any warning gently subsided behind, and then as gently in front, the huge, ugly legs being extended in front of him, and the man signed to me to get off, which I did by getting on his head and letting myself down by a rattan rope upon the driver, who made a step of his back, for even when 'kneeling', as this queer attitude is called, a good ladder is needed for comfortable getting off and on. While the whole arrangement of baskets was being re-rigged, I clambered into a Malay dwelling of the poorer class, and was courteously received and regaled with bananas and buffalo milk. Hospitality is one of the Malay virtues. This house is composed of a front hut and a back hut with a communication. Like all others it is raised to a good height on posts. The uprights are of palm, and the elastic, gridiron floor of split laths of the invaluable *nibong* palm (*oncosperma filamentosum*). The sides are made of neatly split reeds, and the roof, as in all houses, of the dried leaves of the *nipah* palm (*nipa fruticans)* stretched over a high ridge pole and steep rafters of bamboo. I could not see that a single nail had been used in the house. The whole of it is lashed together with rattan. The furniture consists entirely of mats, which cover part of the floor, and are used both for sitting on and sleeping on, and a few small, hard, circular bolsters with embroidered ends. A musket, a spear, some fishing-rods, and a buffalo yoke hung against the wall of the reception-room. In the back room, the province of the women and children, there were an iron pot, a cluster of bananas, and two calabashes. The women wore only *sarongs*, and the children nothing. The men, who were not much clothed, were lounging on the mats....

I had walked on for some distance, and I had to walk back again before I found my elephant. I had been poking about in the scrub in search of some acid fruits, and when I got back to the road, was much surprised to find that my boots were filled with blood, and on looking for the cause I found five small brown leeches, beautifully striped with yellow, firmly attached to my ankles. I had not heard of these pests in Perak, and feared that they were something worse; but the elephant driver, seeing my plight, made some tobacco juice and squirted it over the creatures, when they recoiled in great disgust. Owing to the exercise I was obliged to take, the bites bled for several hours. I do not remember feeling the first puncture. I have now heard that these blood-suckers infest leaves and herbage, and that when they hear the rustling made by man or animal in passing, they stretch themselves to their fullest length, and if they can touch any part of his body or dress they hold on to it, and as quickly as possible reach some spot where they can suck their fill....

Soon the driver jumped off for a gossip and a smoke, leaving the elephant to 'gang his ain gates' for a mile or more, and he turned into the jungle, where he began to rend and tear the trees, and then going to a mud-hole he drew all the water out of it, squirted it with a loud noise over himself and his riders, soaking my clothes with it, and when he turned back to the road again, he several times stopped and seemed to stand on his head by stiffening his proboscis and leaning upon it, and when I hit him with my umbrella he uttered the loudest roar I ever heard. My Malay fellow-rider jumped off and ran back for the driver, on which the panniers came altogether down on my side, and I hung on with difficulty, wondering what other contingencies could occur, always expecting that the beast, which was flourishing his proboscis, would lift me off with it and deposit me in a mud-hole.

On the driver's return I had to dismount again, and this time the elephant was allowed to go and take a proper bath in the river. He threw quantities of water over himself, and took up plenty more with which to cool his sides as he went along. Thick as the wrinkled hide of an elephant looks, a very small insect can draw blood from it, and when left to himself he sagaciously plasters himself with mud to protect himself like the water buffalo. Mounting again, I rode for another two hours, but he crawled about a mile an hour, and seemed to have a steady purpose to lie down. He roared whenever he was asked to go faster, sometimes with a roar of rage, sometimes in angry and sometimes in plaintive remonstrance. The driver got off and walked behind him, and then he stopped altogether. Then the man tried to pull him along by putting a hooked stick in his huge 'flapper', but this produced no other effect than a series of howls; he then got on his head again, after which the brute made a succession of huge stumbles, each one of which threatened to be a fall, and then the driver with a look of despair got off again. Then I made signs that I would get off, but the elephant refused to lie down, and I let myself down his unshapely shoulder by a rattan rope till I could use the mahout's shoulders as steps. The baskets were taken off and left at a house, the elephant was turned loose in the jungle; I walked the remaining miles to Kwala Kangsa, and the driver carried my portmanteau! Such was the comical end of my first elephant ride. I think that altogether I walked about eight miles, and as I was not knocked up, this says a great deal for the climate of Perak. The Malay who came with me told the people here that it was 'a wicked elephant', but I have since been told that it was 'very sick and tired to death', which I hope is the true version of its most obnoxious conduct.

Isabella L. Bird, *The Golden Chersonese and the Way Thither*, John Murray, London, 1883, pp. 297–304.

2
The Jungle is Not Neutral

ARNOT ROBERTSON

Arnot Robertson is the pen name of Eileen Arbuthnot Robertson (1903–61), wife of H. E. (later Sir Henry) Turner, who as General Secretary of the Empire Press Union, travelled widely, often accompanied by his wife, a notable novelist in her day. It was probably in the 1920s that their travels brought them to Malaya and she presumably visited Trengganu, where the novel from which this passage is taken is set. Later she professed 'untrammelled ignorance' of the Malayan jungle, but it is evident that she was sensitive to its sometimes oppressive effect. Much depends on the observer; Colonel Spencer Chapman, in his account of his experiences during the Japanese Occupation period, argued that the jungle is neither hostile nor friendly, and expressed his point in the title of his well-known book, The Jungle is Neutral.

In Arnot Robertson's novel (later adapted as a film) a group of Europeans aboard a coastal steamer discover that the captain is concealing an outbreak of bubonic plague among the Chinese deck passengers. They quit the ship, unobserved, at a minor port (imaginary) on the coast of Trengganu, planning to walk through the jungle to a larger town, where they hope to find another ship to carry them on their journey. The original party is reduced to three but a Malay guide makes four.

THE pitiless sunlight streaming down on us, the fiercely fertile ground over which we made our way, and the clouds of humming winged life into which the dancing air about us steamed as the heat increased—these aspects of the safer, open ground took on for me a new quality of malevolence, toughening my nerves for the dark of the forest, which could be only a little older in iniquity than this conquered ground, where men tormented each other through brief lives.

... I was relieved when the green gloom of the trees closed round us, shutting out the daylight as completely as if dusk had fallen in a few seconds.... Here was a hush against which one could not with dignity put up a defence of human voices, as we had tried to do during the night in the open; it seemed to grow more insistent, heard through our words, and we ceased talking with the same self-consciousness with which Stewart had stopped singing by the stream. Yet in the crowded jungle there could not be absolute quiet but only the precipitate of many soft sounds, making up an oppressive peace. Once that morning a troop of baboons fled whooping overhead, startling us with their shattering din: but there was no room for echo in the forest; the closeness of the foliage muffled all noise, so that the composite stillness of tiny insect hummings and far-off bird calls and the whisper of monstrous growths sucking at the dank earth seemed to leap back afterwards, obliterating the outrage.

The going grew slower and more laborious as we came to soft ground. Even in the drier places where our feet did not sink wearingly, requiring an effort at each step to free them, creepers and thorny tendrils clutched and hindered us. Walking sometimes with a protecting arm before our eyes for a few steps, we stumbled over dead branches and roots hidden in the lush growth underfoot. Ainger looked done-up after we had gone little more than three miles. By then we had been in the jungle for four hours.

Nowhere, in this part of the forest, could we see more than twelve yards round us, and in most cases much less. Our unaccustomed eyes reported all impressions at first as I suppose those of the colour-blind do, in a hundred tones of grey; though our grey was shot with green. Trunk, foliage and creeper in that unhealthy light took on a variety of shades, but one predominant hue which swallowed up all others for a while. Then I saw suddenly, when we had forced our way for some distance into this dim, monotonously coloured world, that the pale-grey mass above me was in reality an immense cluster of hanging yellow flowers busy with golden bees. A great tree, rotten at the core, had fallen in such a way that it would have blocked the path here, where for a hundred yards or more the living roof pressed down so that we walked stooping, and at this point bent double, but the trunk was partially supported still by the mass of its attendant creepers intertwined with others: they had given and sagged but not broken. There was enough room at one side for us to squeeze carefully, for fear of parting the last strand of the sling, whose strength we could not gauge, under this barricade of the jungle's ingenious devising. (All of us but Mrs Mardick came, in time, to believe half consciously in a primeval balefulness, as a kind of vegetable mind behind the seeming-accidental enmity of matter....)

E. Arnot Robertson, *Four Frightened People*, Jonathan Cape, London, 1931, pp. 75–7.

3
A Walk through the Jungle

CARVETH WELLS

Carveth Wells first came to Malaya in 1913 to work as an engineer on the construction of the East Coast railway line, living in jungle camps in Pahang and Kelantan. Owing to ill health he left in 1918, and for the next twenty years he earned his living in the USA as a lecturer and travel writer. At intervals of two years or so, he set off for some comparatively unknown country to gather material for another book and lecture tour, illustrated in later years by his films. In this fashion, after a visit to Japan and China he came back to Malaya late in 1939, for the particular purpose of visiting Theodore Hubback, who was the leading Malayan naturalist and big game hunter at that time.

In the lavish use of capital letters for animal and plant names one can hear the somewhat flamboyant tones of the 'bring 'em back alive' school of lecturer, but it is good description.

TO glance at the map of the Malay Peninsula, it looks quite easy to travel five or ten miles away from a highway; but once you attempt such a journey, it is necessary to have Malay coolies to carry food and equipment and where the jungle is not even traversed by elephant trails, boats must be used. Practically all the main paths through the Malay jungle are ancient or modern elephant trails, some of which I myself surveyed and placed on existing maps. These elephant trails are of course used by many other animals as well as by man. This is known to the leeches which always swarm on such trails, waiting eagerly for a meal. About the length and thickness of a match before it bites you, a leech is as fat as a cigar afterwards.

The Malay land-leech walks like a measuring worm and makes quite rapid progress. In its eagerness for food, the leech tries to enter the eyelets of your boots and suck your blood through your socks. This is a messy business and not very satisfactory to the leech, so that it usually continues to climb up until it reaches your neck. Unfortunately the bite of a clever leech cannot be felt so that you are quite unconscious of the fact that the animal is rapidly filling himself with your blood. This makes him warm so that by the time the leech is gorged and lets go, it often tumbles down your neck without your feeling it. As most people wear wrap-around puttees in the jungle, the leech cannot fall further than your knees, so there it rests and digests its meal. As soon as camp is reached, off come the trousers and out falls the sleepy leech. The next job is to stop the leech bite from bleeding, and this is often quite a problem. Leech-bites often develop into bad sores that take weeks to heal.

This is only one of the many drawbacks to jungle exploration. There are also numerous wasps, from miniature ones that delight in hanging their nests underneath a large leaf in exactly the place you are apt to strike the leaf when walking along a trail, to huge Tebuans. To get a dozen stings from those miniature wasps is not unusual, and although painful, not serious. Quite another affair is the sting of the Tebuan, which is a monstrous wasp with a most ferocious disposition. One sting from a Tebuan has been known to send a Malay to the hospital. Then there are numerous stinging caterpillars. Most of them are hairy and it seems to be the hairs which are poisonous. They have a way of dropping down your neck. If only you had the presence of mind not to squash the animal, all would be well, but usually you feel the caterpillar and immediately swipe at it with your hand. The result is as if someone had seared your skin with a hot iron.

Virgin jungle is comparatively easy to walk through, because the absence of light for centuries has prevented the formation of any dense undergrowth. From outside, the jungle is of very irregular height, the average being about one hundred and twenty-five feet. But dotted about here and there can be seen monster trees over two hundred feet high. The trunks of these trees are usually the only ones visible and they look as if they had been whitewashed. Many of these trees have hanging upon them enormous vines with brightly coloured flowers.

Once inside the jungle, the ground is often covered with masses of beautiful iridescent ferns that change color like silk. There are palms of many varieties but few of them ever grow high enough to show outside the jungle, except the famous Malacca Cane, which is a creeping palm that grows hundreds of feet long.

The main characteristic of a jungle that distinguishes it from an ordinary forest is the extraordinary mixture of trees. In America we are accustomed to forests of Pine, Oak or Cedar; but in a jungle you often have to search to find two trees alike. Jungle trees are usually provided with enormous buttresses to support them in the rotten soil, which is sometimes a hundred feet deep with leaf mold. Some of these jungle giants measure thirty feet across the buttresses, but ten feet from the ground the same tree might only measure eight feet in diameter.

Were it not for their strange calls and noises, you would think that there were no birds in the jungle, for they are only rarely seen. If there are any true song birds in the Malay jungle, I never noticed them. Practically all the birds either sing out of tune or make loud noises or very distinct calls such as those of the Argus Pheasant and the Peacock. The only familiar bird call is the crowing of the Jungle Cock and the clucking of the hen. These lovely little fowl look like the most brightly colored bantams imaginable. I think that the bird most commonly seen because its colours are so conspicuous is the lovely Kingfisher.

Yet the Malay jungle is full of the most spectacular birds in the world, including various kinds of pheasants, wood-pigeon, parakeets, fly-catchers and the giant Rhinoceros Hornbill, which has a habit of walling its mate into a hollow tree at nesting time. One of the most charming sights in the jungle is to see a tree full of Serindits. The Serindit is one of

A female hornbill and her chick, from Alfred Russel Wallace, *The Malay Archipelago, The Land of the Orang-utan, and the Bird of Paradise,* Macmillan & Co., London, 1869.

the favorite birds of the Malays. Until you watch its habits, the Serindit might easily be mistaken for a Love Bird because it is a very similar small green parroquet. But the unusual thing about the Serindit is its habit of sleeping upside down. Sometimes I have come across a dozen of these little birds hanging upside down like fruit upon a tree.

Another strange bird is the Bustard Quail which has several topsy turvy habits. The female is larger than the male; she is pretty but he is ugly. Of course she lays the eggs, but he sits on them and hatches them, and during the mating season the females fight for the males. As for other forms of animal life the jungle swarms with insects from enormous butterflies and moths that measure twelve inches across their outspread wings to cicadae which make trumpet-like noises so loud that towards sundown it is almost impossible to hear yourself speak.

There are elephants, two kinds of rhinoceros, tapir, seladang, bears, three kinds of deer, tigers and over thirty different kinds of monkeys. Over sixty species of bats alone have been discovered, including the huge Flying Fox, which measures five feet across its outstretched wings. Besides the Flying Fox which is the largest bat in the world, the Malay Peninsula claims to have the smallest mammal (in weight) in the world, which is also the smallest bat.

Other flying animals of the Malay jungle are the Flying Lemur, Flying Lizard and Wallace's Flying Frog which can skim through the air on the soles of its enormous webbed feet. Crocodiles are numerous, especially on the lower reaches of the rivers which run into the Straits of Malacca, and they take a larger toll of human life than any other animal. Specimens thirty feet long have been recorded but those exceeding sixteen feet are exceptional. The huge Reticulated Python is the largest Malayan snake, growing to a length of thirty feet, but the most dangerous snake in Malaya and in the world for that matter is the Hamadryad or King Cobra, the largest of all poisonous snakes. This formidable snake does not hesitate to attack a man. It always seems to be in a bad temper. Specimens measuring eighteen feet six inches have been recorded, and when it is remembered that a snake can strike half its own length, some idea can be had of the great danger there is from this reptile. Unfortunately the Hamadryad is by no means an uncommon snake.

It is undoubtedly on account of the many dangers of the jungle that the Malays themselves keep out of it except on business. If a Malay wants some jungle produce such as rattan or damar or palm leaves, he first mutters a few prayers to the spirits of the jungle, explaining the purpose of his visit.

Carveth Wells, *North of Singapore*, Robert M. McBride and Co., New York, 1940, pp. 149–53.

4
The Flight of the Aborigines

HUGH CLIFFORD

Hugh Clifford first came to Malaya in 1883 at the age of seventeen. After a few years in Perak under Hugh Low (Passage 36) and Frank Swettenham (Passage 67) as Residents, Clifford was sent to the capital of Pahang in 1887 as British consular representative to what was still an independent State. Pahang came under British control in 1889 and Clifford (Passages 7, 9, 65, 66, and 85) spent an eventful decade there until he left Malaya (to pursue a successful career in other colonies, mainly in West Africa); he returned to Singapore as Governor and High Commissioner in 1927, but retired on health grounds in 1929 and died in 1941. Most of his imaginative and vivid writing is description, sometimes semi-fiction, of his experiences in Pahang in the 1890s.

In the nineteenth century the aborigine (Sakai) peoples, who lived in the remote parts of the Malayan jungles, traded with Malay villages but risked being captured and enslaved by the villagers. Clifford here describes a desperate—and in the end unsuccessful—attempt to throw pursuers off their track.

THE Malay Peninsula is one of the most lavishly watered lands in all the earth. In the interior it is not easy to go in any direction for a distance of half a mile without encountering running water, and up among the foothills of the main range, when navigable rivers have been left behind, travelling through the forest resolves itself into a trudge up the valleys of successive streams, varied by occasional scrambles over ridges of hill or spurs of mountain which divide one river system from another. Often the bed of the river itself is the only available path, but as wading is a very fatiguing business, if unduly prolonged, the banks are resorted to whenever a game-track or the thinning of the underwood renders progress along them practicable.

The Sakai fugitives, however, did not dare to set foot upon the land when once they had quitted their camp, for their solitary chance of throwing pursuers off their track lay in leaving no trace behind them of the direction which they had followed. Accordingly they began by walking up the bed of the little brawling torrent, swollen and muddy from the rains of the previous afternoon, and when presently its point of junction with a tributary stream was reached, they waded up the latter because of the two it seemed the less likely to be selected. It was miserable work, for the water was icy cold, and the rivulet's course was strewn with ragged rocks and hampered by fallen timber; but the Sakai seemed to melt through all obstructions, so swift and noiseless was their going. They crept through incredibly narrow places; they scrambled over piles of fallen timber without disturbing a twig or apparently leaving a trace; and they kept strictly to the bed of the stream, scrupulously avoiding even the brushwood on the banks and overhanging branches, lest a broken leaf should betray them to their pursuers.

The men carried their weapons and most of their few and poor possessions; and the women toiled along, their backs bowed beneath the burden of their rattan knapsacks, in which babies and carved receptacles made of lengths of bamboo jostled rude cooking-pots of the same material and scraps of evil-looking food. Children of more than two years fended for themselves, following deftly in the footsteps of their elders, many of them even helping to carry the property of the tribe.

The long procession wound its way in single file up the bed of the tributary stream until the midday sun showed clearly their heads through the network of vegetation. The Sakai all walked in precisely the same manner, each foot being placed exactly in front of its fellow, and each individual treading as nearly as possible in the footsteps of the man in front of him. Experience must, in some remote and forgotten past, have taught the forest-dwellers that this is the best and quickest way of threading a path through dense jungle, and in the course of time experience has become crystalized into an instinct, so that to-day, even when walking along a broad highway, the Sakai still adopt this peculiar gait. You may mark a similar trick of successively placing the feet one exactly in front of the other in many wild animals whose lives have been passed in heavy forest.

At last old Ka', who was leading, halted, and his followers stood in their tracks while he grunted out his orders. A steep hill, some five hundred feet high, rose abruptly on their right. It was covered with jungle through which the eye could not penetrate in any place for more than a few yards; but all the Sakai knew that its crest was a long spur or hogs-back, which if followed for a matter of half a mile would enable them to pass down into the valley of a stream that belonged to a wholly different river system. By making their way up its bed they in time would win to the mountains separating Perak from Pahang; and when the raiders, if they succeeded in picking up the carefully veiled trail, found that the fugitives had gone so far, it was possible that they might be discouraged from further pursuit, and might turn their attentions to some more accessible band of wandering Sakai. The first thing, however, was to conceal all traces of the route which Ka's party had taken, and he therefore bade his people disperse, breaking up into little knots of two or three, so that no definite, well-defined trail might be left as a guide to the pursuers. Later the tribe would reassemble at a spot appointed by him. The Sakai were well versed in all such tricks, and very few words and no explanations were needed to convey to them an understanding of their leader's plan. In the space of a few seconds the little band of aborigines had broken up and vanished into the forest as swiftly and silently as a bank of mist is dispersed by a gust of morning wind.

Hugh Clifford, *The Further Side of Silence*, Doubleday, Page & Company, New York, 1927, pp. 253–7.

5
Going Up Gunong Tahan—and Down Again

WALTER SKEAT

Walter Skeat served as an administrator in Selangor from 1891 to 1899, when he became joint leader of a scientific expedition to the Malay States of the north-east (including those in Siam), of which little was then known. In this letter Skeat describes how he made an ill-prepared attempt to climb Gunong Tahan, the highest peak in Malaya, and narrowly escaped disaster and loss of life. The hardships broke Skeat's health and he was unable to continue his career in Malaya, although he lived on until 1953.

Becher's death, to which Skeat refers, occurred during a previous attempt on the mountain in 1893. Apart from the uncertain weather and rivers in sudden flood, the main obstacle is a series of rock-faces, for which the first successful party to climb to the peak in 1906 brought light scaling ladders.

I have just got back safe though half-starved from Gunong Tahan and write to let you know that you need no longer be anxious on that account. I found that the rains had set in already, and set in so heavily, on the mountain that I think it would be altogether too risky to try and get the men up there just now. In fact I do not believe it would be feasible, as the Tahan was three times in flood while we were on the range. I look upon it as little short of a miracle that we were not caught by them.

I was told by a Malay who saw Beecher (*sic*) drowned that he was swept off a very high rock, which nobody thought would be reached, by a flood which came roaring down the valley by day when there was no rain. He had about 80 men with him. I had only 6 men with me and was my own guide, making a time and compass survey while I carried the tent as we climbed.

We could barely carry enough rice to last and carrying meat and vegetables was out of the question. So for 4 weeks I ate rice with the men, until the rice ran short and the matches gave out. We finished up by living like the wild tribes on such fruits as we could find in the jungle.

I went to make a personal reconnaissance, as I told Shipley, intending to go up a few high crags, and try to locate the mountain, which is more than has ever been done by previous expeditions. But the second crag that we climbed led up to eight successive peaks at least, each a little higher than the other. Finally we reached the side of a big mountain which we took to be an eastern peak of a gigantic mass of mountain which blocked up the end of the valley. The peak we climbed to within about 50 yards of its highest 'aiguille', so far as I could judge. We found from enquiries, when we got back to civilisation, among the Malays who had been up the valley with former expeditions, that what we had been climbing was really Gunong Tahan (Mount Tahan), the west peak, which appeared to be a little lower, being Gunong Larong.

We had to contend with hunger. I might, I think, without exaggeration call it slow starvation, wintry cold at night, when we slept for weeks in our rain-sodden and at last almost rotting clothes, tropical showers by day, which chilled one to the bone, ragged knife edges on the ridge, bare granite precipices—in many case actually over-hanging at the top—and jungle that was in many places thick set, with thorny trees and shrubs that tore our clothes (and my topee) to ribbons. Then the swamp water, through which we had to go, continually poisoned the skin of our feet—the men's not badly but mine became so swollen and painful that it was agony to walk in boots. I had to take mine off and wrap up my feet on the way down, the jungle being so full of thorns that I could not go bare-foot. Then one of the men got ill and I had to doctor him.

By good luck we had now got to a stream where we stopped to make rafts. But we had only gone down a few bends when we had to abandon the rafts, as we came to a continuous chain of rapids over half a mile long. It was a little better here. We got past the rapids and built fresh rafts. Then we went downstream till we came back to the Tahan again. After 2 or 3 days hard labour at the rapids, which are far worse and more numerous on this than on any other river, we got to its juncture with the Tembeling and *back to life* again.

Still we had 6 days' journey to get back. When we got back to the watershed, Dollah, the man who was so ill before, had to be carried through the jungle for a day and a half. Just after we had started, with three men who *professedly* knew the way, we all got lost. By the time I had found the way again for them one of the other men, who had been snake-bitten and cauterised by me as we came down the hill, got a sharp attack of ague, which necessitated my carrying his pack for that day and the next.

I have not yet mentioned that a third developed fever and a fourth also was ill. Meanwhile we were living upon a cupful or two of uncooked rice and wild fruits.

Still we got through all right. What is more we have been at least nearer the summit of that hill than any man who ever lived, and were only beaten by a neck so to speak at the last. I hear now that at least two expeditions have tried the mountain from here but neither by this nor the Tahan route has any living man yet reached the mountain itself, which is the centre of a vast cloud of superstition only less dense than the blind and baffling fogs that checked us daily as we went up it.

To the Malays it is the abode of all the most malignant spirits, and contains the matrix of all gold and silver ores, which is guarded by gigantic man-eating apes as big as bulls and immense serpents and mosquitoes as big as fowls.

The very day I got back we secured from the people here a young tapir. While I was absent on the Tahan range Evans found several more specimens of Peripatus.

Walter Skeat, letter of 15 September 1899, *Journal of the Malaysian Branch of the Royal Asiatic Society*, Vol. LXI, Pt. 1, 1988, pp. 142–3.

On the Waters

6
By Boat to Kuala Langat

EMILY INNES

Emily Innes was the wife of James Innes, who held administrative posts in Sarawak, Selangor, and Perak between 1871 and 1882. When he left Sarawak and took up the post of Collector (District Officer) at Kuala Langat, the royal capital of Selangor, in mid-1876, Emily joined him at his new station. This passage describes the final stage of her journey, from Klang to Kuala Langat by boat, to her new home.

Emily never met Isabella Bird (see Passage 1 above) but she read her book, when it appeared in 1883. Emily accepted that Miss Bird's account of her brief tour, as the guest of senior officials, was 'perfectly and literally true', but Emily felt strongly (she was a woman of very definite views) that it completely failed to depict the monotony and hardship experienced by those who, like herself, lived in remote parts of the Malay States for years. So she wrote her own book to describe her six years (1876–82) in the Malay States. Despite its bitterness we may accept that what she wrote is an accurate—and also a vivid—description. Although it is personal and not very penetrating, it is one of the best pictures of European (and Malay) life in western Malaya around 1880.

ARRANGEMENTS having been made for sending me on, I started in a native boat on the Langat river for Langat. This was my first experience of a long voyage in a native boat; it was so uncomfortable that I resolved it should be my last, and during my six years' stay in the Straits I kept to my resolution, thus losing many an expedition which might have varied the monotony of my life.

The discomfort was not the fault of the Resident, who had done what he could for me, but that of the boat, which reminded me of the oubliettes that I had seen in the dungeons of the Inquisition at Rome: it was impossible to sit or stand in it at all, on account of the very low palm-leaf awning; one could only lie down in it, and that in a very uncomfortable position. The Resident had provided a boat-mattress of the usual sort, about an inch thick and very hard, plenty of food, and an old woman.

The use of the old woman, he explained, was to give me importance in the eyes of the natives, who would think it very *infra dig.* were I to travel without any female attendant. Barring the honour, I would gladly have dispensed with this old creature's presence: she was frowzy in her garments and very dirty in her habits; she chewed betel constantly, and her talk required Bowdlerizing so much that I soon pretended to know no Malay, and thus tried to silence her.

The awning of the boat was so low at the sides that it was impossible even to see the banks of the river. In front it was a

A river boat of the kind in which Emily Innes and also Isabella Bird (Passage 8) made their journeys, from Major Fred McNair, *Perak and the Malays: 'Sarong and Keris'*, Tinsley Brothers, London, 1878.

little higher, but the prospect so afforded was not exhilarating, as it consisted only of eight brown flat noses, eight enormous mouths, and eight pairs of greasy brown shoulders, the property of the boatmen, each of whom was more hideously ape-like than his neighbour. They had very little clothing on, but quite enough to taint the air with indescribable native bouquet, in which cocoanut-oil is the chief ingredient. Books I had none; of mosquitoes there were thousands, and of ants many hundreds, which swarmed over the food and my clothing and completed my disgust.

The voyage ought properly to have lasted eight hours, in which case it would have been endurable, but Malays never hurry themselves when there is no white man present, and these boatmen consequently took it very coolly. Whenever they passed a hut on the bank—which, fortunately, was not often, for the country is very thinly populated—they stopped and went on shore, returning with hands full of bananas and sugar-cane, begged, borrowed, or stolen from the owners. In this way they contrived to lose the tide; this exactly suited their views, as they now had an excuse for sleeping several hours until it turned. The result was that the voyage took twenty-four hours instead of eight. I was in perfect misery the whole time, for, besides the ants, smells, mosquitoes, dirty old woman and uncomfortable position before mentioned, it was a great trial to be deprived of the three daily baths necessary to keep one cool in this climate.

For an English *man*, a long boat-voyage is not nearly so bad, as he can bathe whenever he likes. He has only to slip on a Malay sarong and bathe in public, as the natives do. My old duenna did so, standing in her sarong, while a friendly boatman poured buckets of water over her head, and the whole population of the boat and wigwam looked on with interest. She pressed me to imitate her example; but such is the force of early prejudice, that, finding privacy out of the question, I preferred to endure my sufferings like a martyr. A sarong, perhaps I should explain, is a piece of drapery like a 'round towel' in form, which, after the first bucket of water, clings to the figure like the drapery of a Greek statue. My elderly companion did not resemble any Greek statue I had ever seen, being extremely stout and repulsive-looking.

A weary afternoon, evening, and night, succeeded each other, and were succeeded in their turn by a still wearier morning and afternoon. I constantly asked the boatmen how much farther it was to Langat, but they did not appear to know exactly. Sometimes they told me it was so many tanjongs (bends of the river) off; but when I had counted that number go by, they said they had been mistaken, and it was more tanjongs still. If I asked how many more? I reeived the unfailing Malay answer. 'Tidak tuntu'—'It is not certain.' At last, when in despair I had given up asking them, I was informed Langat was in sight. Soon we arrived at the landing-place before mentioned, and having reached terra firma as well as limbs stiffened by twenty-four hours of oubliette would allow, I walked towards the house. Mr Innes had had no notice of my coming, as postal arrangements there were none in the country, but a boatman ran on ahead to tell him, and he came out to meet us.

The house was worse than I had expected....

Emily Innes, *The Chersonese with the Gilding Off*, 2 vols., Richard Bentley and Son, London, 1885, Vol. 1, pp. 10–15.

7
Coming Home to a Strange Land

HUGH CLIFFORD

In 1902 Clifford, then on leave in London, was deputed to escort Sultan Idris of Perak, who had come over to attend the Coronation of King Edward VII. In the course of this visit Clifford witnessed a very fraught encounter between the Sultan and his son (the future Sultan Iskander), who had been sent to study in England at the age of fourteen and had not seen his father for five years, during which time the young man had become very Western in his ways. This episode prompted Clifford to write a novel concerned with the problems in England, and then in Malaya on his return, facing a young Malay after a long period of absence at an impressionable age.

To conceal the source of his story, Clifford sets it in Pahang ('Pelesu'), not Perak. He here describes a journey by boat through the estuary of the Pahang River to the state capital, in which Raja Saleh, returning after a long absence, has his first encounter with a home environment which has become strange to him.

THE arrival at the mouth of the Pelesu River and the ten miles' journey upstream to his father's capital caused a mighty vibration of the chords of memory. The villages under the shady cocoanut palms, the deep fringes of restless casuarina trees with each delicate spine atwitter in the breeze, the long yellow sand-spits, the thick-set wooden light-house painted black and white and squatting squarely on its four sturdy legs, the irregular, creamy line of the bar where the waters of the river contended eternally with the tides—every one of these things cried its separate welcome to Saleh. Then came the noisy transfer from the yacht to the native boat with its crowd of gayly-clad Malays saluting him with lifted paddles and curious, interested faces, and a somewhat ignominious crawl into the wooden, palm-roofed shelter at the stern.

Saleh seated himself on the carpet which covered the deck of this cabin, and noted with interest the spears slung from thongs from the roof. With a shock of wonder he found him-self recollecting not only each spear, but the individual nick-name that it bore, and something even of its fabled record! The very existence of these famous weapons, he was certain, had not been so much as remembered by him for years. Where, then, he asked himself, had all this lore, that now recurred to him so readily, been hidden that long while? And yonder, upstream near the bend, lay an island—a tiny piece of earth supporting a dozen cocoanut palms and a hut—Pulau Kapas—Cotton Island—though there was not a cotton tree within a mile of it. How was it that the name, so long unthought upon, leaped now unbidden to his lips? And it was the same with everything—people, places, things—he re-membered each one of them vividly, though the faces were older, the dimensions of inanimate objects had shrunk curi-ously, and with them came rags and tags of story which he could never even remember to have heard. Each one of the four grave and aged headmen who squatted about the cabin door in silence, their attitudes submissive, their hands clasped in their laps, their backs to the straining men at the paddles, was known to him with an intimacy that was startling, since no thought of them had so much as crossed his mind for years.

There was to Saleh something vaguely disquieting, even terrifying, about this sudden sharp assertion of the powers of unsuspected memory. It seemed to him as though he were the possessor of a dual personality, and that one of the egos within him had been long lapped in slumber and now was abruptly awakened. It was playing queer and uncanny tricks upon him already. It was strange to him; he did not know where to have it, what to expect from it. It made him conscious of an extraordinary feeling of uncertainty, of insecurity about himself.

With the awkwardness bred of long desuetude and emphasized by the fact that he was still clad in European fashion, Saleh sat cross-legged on the carpet just within the doorway of the tiny cabin, watching the familiar landmarks come up one by one, each in its turn to drop behind as the boat leaped forward to the ordered thump and splash and rhythm of the paddles. Above him hung the historic spears: before him sat the grave courtiers, dressed in correct Malay costume, *kris* [daggers] stuck in their girdles, twisted handkerchiefs on their heads, their faces immovable as though carved out of mahogany. Over their shoulders Saleh could see the bright silks of forty paddlers in kaleidoscopic movement, as the shining paddle blades, dripping gems of the sunlit water, rose and fell. Beyond them again the long reach of the river, flanked by villages, set with islands, busy with the traffic of small craft, was visible in swiftly changing glimpses. The steersman, perched in the *magun* [deck-house] on the roof of the cabin, lifted up his voice and began to keen a boat song, an elusive, plaintive melody pitched in a minor key, instinct with the unresisting, patient melancholy of his race, and as the men at the paddles took up the refrain, Saleh felt as though the very heart-strings of his soul were being made the instrument of that music. A little puff of scent-laden wind wandered down the river and blew upon his cheek. It was to him as though the land that gave him birth was greeting him with a kiss. Yes, yes, yes, it was here, here that he belonged!

At the landing-stage which ran out into the water near the centre of the Kampong Raja—the King's Compound—a long string of ramshackle buildings, each in its own ill-kept grounds, the whole surrounded by a high fence of split bamboos—a big crowd of natives had congregated to witness the arrival of their Sultan's son. As Saleh stepped ashore every soul present squatted suddenly, and as he stood still in momentary surprise, those who were following him immediately imitated their example. Saleh found himself in an instant awfully alone—the only erect figure in that wide multitude, every eye in which was fixed upon him. He felt himself flush painfully under his dark skin.

Hugh Clifford, *A Prince of Malaya*, Harper & Brothers, New York, 1926, pp. 131–3; reprinted from 'Sally', *Blackwoods Magazine*, November 1903–February 1904, and *Saleh: A Sequel*, Blackwood, London, 1908.

8
Along the Linggi River by Boat

ISABELLA BIRD

In the course of her brief visit to the western Malay States in 1879 (see Passage 1 above), Isabella Bird made a boat journey up the Linggi River to Seremban, in the company of the Misses Shaw, two daughters of the Resident Councillor, Malacca, whose frailties were exposed by a gruelling trip without earning them much sympathy from Isabella, a tough veteran of such experiences.

WE left Malacca at seven this morning in the small, unseaworthy, untrustworthy, unrigged steam-launch *Moosmee*, and after crawling for some hours at a speed of about five miles an hour along brown and yellow shores with a broad, dark belt of palms above them, we left the waveless, burning sea behind, and after a few miles of tortuous steaming through the mangrove swamps of the Linggi river, landed here to wait for sufficient water for the rest of our journey.

This is a promontory covered with coco-palms, bananas, and small jungle growths. On either side are small rivers densely bordered by mangrove swamps. The first sight of a real mangrove swamp is an event. This *mangi-mangi* of the Malays (the *Rhizophera mangil* of botanists) has no beauty. All along this coast within access of tidal waters there is a belt of it many miles in breadth, dense, impenetrable, from forty to fifty feet high, as nearly level as may be, and of a dark, dull green. At low water the mangroves are seen standing close packed along the shallow and muddy shores on cradles or erections of their own roots five or six feet high, but when these are covered at high tide they appear to be growing out of the water. They send down roots from their branches, and all too quickly cover a large space. Crabs and other shellfish attach themselves to them, and aquatic birds haunt their slimy shades. They form huge breeding grounds for alligators and mosquitoes, and usually for malarial fevers, but from the latter the Peninsula is very free. The seeds germinate while still attached to the branch. A long root pierces the covering and grows rapidly downwards from the heavy end of the fruit, which arrangement secures that when the fruit falls off the root shall at once become embedded in the mud. Nature has taken abundant trouble to ensure the propagation of this tree, nearly worthless as timber. Strange to say, its fruit is sweet and eatable, and from the fermented juice wine can be made. The mangrove swamp is to me an evil mystery.

Behind, the jungle stretches out—who can say how far, for no European has ever penetrated it?—and out of it rise, jungle-covered, the Rumbow hills. The elephant, the rhinoceros, the royal tiger, the black panther, the boar, the leopard, and many other beasts, roam in its tangled, twilight depths, but in this fierce heat they must all be asleep in their lairs. The Argus-pheasant too, one of the loveliest birds of a region whose islands are the home of the Bird of Paradise, haunts the shade, and the shade alone. In the jungle too is the beautiful bantam fowl, the possible progenitor of all that useful race. The cobra, the python (?), the boa-constrictor, the viper, and at least fourteen other ophidians, are winding their loathsome and lissome forms through slimy jungle recessses; and large and small apes and monkeys, flying foxes, iguanas, lizards, peacocks, frogs, turtles, tortoises, alligators, besides tapirs, rarely seen, and the palandok or chevrotin, the hog deer, the spotted deer, and the sambre [sambir], may not be far off. I think that this part of the country, intersected by small, shallow, muddy rivers, running up through slimy mangrove swamps into a vast and impenetrable jungle, must be like many parts of Western Africa.

One cannot walk three hundred yards from the station, for there are no tracks. We are beyond the little territory of Malacca, but this bit of land was ceded to England after the 'Malay disturbances' in 1875, and on it has been placed the Sempang police station, a four-roomed shelter, roofed with *attap*, a thatch made of the fronds of the *nipah* palm, supported on high posts,—an idea perhaps borrowed from the mangrove,—and reached by a ladder. In this four Malay policemen and a corporal have dwelt for three years to keep down piracy. 'Piracy', by which these rivers are said to be infested, is a very ugly word, suggestive of ugly deeds, bloody attacks, black flags, and no quarter; but here it meant, in our use of the word at least, a particular mode of raising revenue, and no boat could go up or down the Linggi without paying black mail to one or more river rajahs.

Our wretched little launch, moored to a coco-palm, flies a blue ensign, and the Malay policemen wear an imperial crown upon their caps,—both representing somewhat touchingly in this equatorial jungle the might of the small island lying far off amidst the fogs of the northern seas, and in this instance at least not her might only, but the security and justice of her rule.

Two or three canoes hollowed out of tree trunks have gone up and down the river since we landed, each of the inward bound being paddled by four men, who ply their paddles facing forwards, which always has an aboriginal look, those going down being propelled by single, square sails made of very coarse matting. It is very hot and silent. The only sounds are the rustle of the palm fronds and the sharp din of the cicada, abruptly ceasing at intervals.

In this primitive police station the notices are in both Tamil and Arabic, but the reports are written in Arabic only. Soon after we sat down to drink fresh coco-nut milk, the great beverage of the country, a Malay bounded up the ladder and passed through us with the most rapid and feline movements I have ever seen in a man. His large, prominent eyes were fixed, tiger-like, on a rifle which hung on the wall, at which he darted, clutched it, and, with a feline leap, sprang through us again. I have heard much of *amok* running lately, and have even seen the two-pronged fork which was used for pinning a desperate *amok* runner to the wall, so that for a second I thought that this Malay was 'running amuck'; but he ran down towards Mr Hayward, our escort, and I ran after him, just in time to see a large alligator plunge from the bank into the water. Mr Hayward took a steady aim at the remaining one, and hit him, when he sprang partly up as if badly wounded, and then plunged into the river after his companion, staining the muddy water with his blood for some distance.

Police Station, Permatang Pasir. Sungei Ujong, 5 P.M.—We are now in a native state, the territory of the friendly Datu Klana, Syed Abdulrahman, and the policemen wear on their caps not an imperial crown, but a crescent, with a star between its horns.

This is a far more adventurous expedition than we expected. Things are not going altogether as straight as could be desired, considering we have the Governor's daughters with us, who, besides being very precious, are utterly unseasoned and inexperienced travellers, quite unfit for 'roughing it'. For one thing, it turns out to be an absolute necessity for us to be out all night, which I am sorry for, as one of the girls is suffering from the effects of exposure to the intense heat of the sun.

We left Sempang at two, the Misses Shaw reeling rather than walking to the launch. I cannot imagine what the mercury was in the sun, but the copper sheathing of the gunwale was too hot to be touched. Above Sempang the river narrows and shoals rapidly, and we had to crawl, taking soundings incessantly, and occasionally dragging heavily over mud banks. We saw a large alligator sleeping in the sun on the mud, with a mouth, I should think, a third of the length of his body; and as he did not wake as we panted past him, a rifle was loaded and we backed up close to him; but Babu, who had the weapon, and had looked quite swaggering and belligerent so long as it was unloaded, was too frightened to fire, the saurian awoke, and his hideous form and corrugated hide plunged into the water so close under the stern as to splash us. After this alligators were so common, singly or in groups or in families, that they ceased to be exciting. It is very difficult for anything to produce continuous excitement under this fierce sun, and conversation, which had been flagging before noon, ceased altogether. It was awfully hot in the launch, between fire and boiler heat and solar fury. I tried to keep cool by thinking of Mull, and powdery snow and frosty stars, but it would not do. It was a solemn afternoon, as the white, unwinking sun looked down upon our silent party, on the narrow turbid river, silent too, except for the occasional plunge of an alligator or other water monster,—on mangrove swamps and *nipah* palms dense along the river side, on the blue gleam of countless kingfishers, on slimy creeks arched over to within a few feet of their surface by grand trees with festoons of lianas, on an infinite variety of foliage, on an abundance of slender shafted palms, on great fruits brilliantly coloured, on wonderful flowers on the trees, on the *hoya carnosa* and other waxen-leaved trailers matting the forest together and hanging down in great festoons, the fiery tropic sunblaze stimulating all this over production into perennial activity, and vivifying the very mud itself.

Occasionally we passed a canoe with a 'savage' crouching in it fishing, but we saw no other trace of man, till an hour ago we came upon large coco groves, a considerable clearing in the jungle, and a very large Malayan-Chinese village with mosques, one on either side of the river, houses built on platforms over the water, large and small native boats covered and thatched with *attap*, roofed platforms on stilts answering the purpose of piers, bathing-houses on stilts carefully secluded, all forming the (relatively) important village of Permatang Pasir.

Isabella L. Bird, *The Golden Chersonese and the Way Thither*, John Murray, London, 1883, pp. 162–7.

9
Down the Pahang River to a New World

HUGH CLIFFORD

As in Passage 7, Clifford depicts a situation of social change in the Malay States around the end of the nineteenth century. The formal abolition of 'debt-bondage', a traditional Malay institution which in British eyes was slavery, was not immediately effective in remoter areas. This passage describes the escape from bondage in the interior of Pahang of two Malay girls, aged nine and eleven, and their fear of the wider world in which they hope to find refuge. The story is no doubt based upon an actual episode known to Clifford; it ends happily for the two girls with the thundering humiliation of Che' Awang, their former master.

MINAH had managed to hide a couple of boat-paddles together with her other gear, and each child took one of them, Minah steering, while Iang paddled at the bow, still sobbing miserably.

Like all Malayan children, born and bred on the banks of the rivers, Iang and Minah were as much at home upon the water as they were upon the dry land. They both understood the management of a dugout as thoroughly as one could desire; they both swam with ease and grace; and for them the river itself had no terrors. What they did fear most mightily was the Unknown into which they were journeying. The section of the world with which the up-country native is familiar is often very circumscribed indeed. . . .

The world for Minah and her sister had hitherto been bounded by a village half a mile up-stream, whither they had gone occasionally to help in the weeding of the crops; by the little shady graveyard, where the round headstones stood in disorder among the rank grasses and the *sudu* [*euphorbia*] plants; by half a mile of jungle beyond the rice swamps at the rear of the compound; and across the river by half a mile of virgin forest which rose sheer from the banks of the stream. Now they were setting out upon a journey which they knew would last for several days. They had no scale of distance by which to measure things other than the fathom which they now felt to be alarmingly inadequate to their requirements—and indeed, when you try to reduce eighty miles of running water to fathoms, the figures that result are somewhat terrifying. Not that the little girls attempted to do anything of the sort. The very fact that the distance ahead of them was measureless added not a little to its terrors. If you can put yourself in the position of a traveller who, on setting forth, finds that miles, as units of distance, have shrunk into utter insignificance, as a man might do who was starting upon a voyage to one of the more remote stars, you will be able to understand dimly in what guise the prospect of this journey presented itself to the minds of these little brown babies, and you will perhaps sympathise with Minah, who soon found her courage oozing out of the tips of her fingers and the tears running down her cheeks. As for Iang, she had been weeping piteously from the first, and every now and again she implored her sister to return and abandon the enterprise. But Minah, though she wept furtively in the darkness, and was afraid to the marrow of her little bones, would not allow her resolution to be shaken. She whispered words of encouragement to Iang, steadying her voice bravely for the purpose; and thus in tears and in dread the long journey was begun.

The musical noises of the jungle night made sleepy melody for the children as they journeyed. The bell-like note of the tree-frog, the hoot of the peafowls, the ticking of the insects, the very distant trumpeting of an elephant, the sharp bark of a stag near the edge of a clearing on the bank, the snorting of wild swine heard once or twice as the passage of the boat near the shore startled a herd of them into panic-stricken flight, and once the brisk and angry clang of a gun—each came in turn, emphasising the noisy silence of the forest. All night the children paddled bravely, falling asleep over their work, and recovering themselves with a start just as their paddles were dropping from their grasp. They bent forward, borne by the current rather than by the strength of their feeble rowing—now grounding upon a sand-bank, to get clear of which called for a heart-breaking struggle; now wandering into a backwater, which they mistook for the main stream; while ever and anon a rock would start up before them out of the gloom menacingly, as the roar of the waters eddying around it set their little hearts beating with terror.

And ever the agony of the effort by which alone they could keep awake, and the utter exhaustion that exertion brought them after a day of such unusual emotions, weighed heavily like a tangible burden. They seemed to be part and parcel of a hideous nightmare: fear of the darkness oppressed and daunted them; dread of the journey that lay ahead was magnified now exceedingly; pursuit, capture, and dire punishment seemed to be their certain fate; and if they did succeed in reaching the town that men called Pekan, their only hope of deliverance lay in the protection that might be afforded to them by the white strangers—a race of men as weird and awful to these little brown girls as the ogres of our own nursery tales. Even Minah lost heart completely now; and had it not been that the current forbade retreat, she would very willingly have returned to the familiar village, and have submitted to the worst that Che' Awang could inflict.

The dawn, breaking wanly, looked into the faces of the two children through a drenching veil of mist, and found them grey and haggard, their figures bent and drooping with fatigue. Minah, chilled to the bone and sodden with the damp dews, made shift to guide the dugout into a tiny creek over which the jungle trees arched, forming a dark and gloomy tunnel. Here the boat was hidden, and the two children dragged their cramped limbs on to the shore, and fell asleep as soon as their heads touched the ground. There they lay, locked in one another's arms, utterly alone in the great forest.

It was past noon when they woke, and Minah, resolute again now that daylight had brought a renewal of her courage, washed the rice in the stream, and, after a mighty tussle with a tinder box, made a fire upon which to cook it. The food refreshed them, and even Iang began to take a more hopeful view of their prospects. After they had eaten, they slept again; and when they woke, the jungle was noisy with its evensong, and the darkness was beginning to fall. They ate the remainder of their boiled rice, packed their gear into the dugout, and set forth anew on their adventurous way.

This night passed like the preceding one, but it held something less of terror, for their unexpected success had given the children confidence. All the next day they lay hidden in the jungle, and once or twice they saw a boat creep by on its way up-stream, or speed swiftly down, borne by the current.

In one they thought they recognised the figures of Che' Awang himself and a few of the villagers with whom they were familiar; and Iang, gripped suddenly by a keen home-sickness, was hardly to be restrained from shouting to them to attract attention. Even hostile faces, so be it they were not those of strangers, seemed welcome to her in the heart of this vast and fearful wilderness into which her little sister had led her. Minah prevented the indiscretion just in time, and again the journey was continued as soon as the night had come.

For more than a week the two little girls travelled in this fashion, lying up in the forest by day, and speeding forward as best they might under the screen of darkness; and as the dawn was yellowing for the ninth day, they found themselves in sight of the largest 'compound' that they had ever seen. The river was nearly a mile across at this place, so that more than once the children had questioned whether they had not already reached the sea—the mighty waters, 'bigger than a river', of which they had heard men speak. The banks were covered with villages as far as the eye could carry; the very islands spattered over the broad reaches were thick with palm groves and the thatched roofs of houses; and far away there rose certain huge white objects covered by great expanses of red stuff, or by things which looked like giant kerosene-tins. These were the whitewashed stone and brick shops of the Chinese, with their roofs of tile or corrugated iron—wonderful things in the eyes of the little up-country savages.

'O Minah,' whispered Iang, frightened out of her wits, 'let us make for the jungle.'

The jungle is always the place of refuge in which the natives of the interior seek safety in time of peril.

'Nay,' cried Minah, biting her lips resolutely, 'This must be Pekan.'

Borne down by the sheer strength of her small sister's will, Iang said no more; but she gazed at the river-side town—a poor, shabby, little place, in all truth, but marvellous and awe-inspiring to her—with terror in her heart.

Hugh Clifford, 'Two Little Slave Girls', *Malayan Monochromes*, John Murray, London, 1913, pp.138–44.

10
A Voyage to Port Dickson

ETHEL DOUGLAS HUME

In 1899 Ethel Douglas Hume (later Mrs H. Thomson) came to spend a few months in Malaya with her brother, a government official then living in Kuala Lumpur. She was invited to join friends at the seaside at Port Dickson but there was no rail link between Selangor and Negri Sembilan, and so she went down to Malacca and, as here recounted, embarked on a coastal steamer to complete her journey to Port Dickson.

Her book, published in 1907, is an entertaining account of her travels, her stay in Malaya, and of her later travels to Ceylon, Japan, and China.

THE steamer on which I recommenced my roundabout route to Port Dickson was known as the overland vessel. One morning the passengers had awakened to find the boat stationary, her nose run up on to land, with her prow heading straight for a lighthouse in the most confiding way.... I embarked with difficulty over lower decks covered with pigs in their pokes, and upper decks littered from stem to stern with prostrate Chinese who, fortunately, seemed to have no objection to being trodden upon. I trusted I should not be expected to take the same recumbent position, and was relieved when the Scotch captain invited me on to the bridge, and squeezed me into the only available corner under the wheel....

The captain's conversation was presently interrupted by some Malay ladies who spied me out and came to me, stepping over the prostrate Chinese with the vigorous arm swinging walk of the feminine Malay. Their eloquence almost outvied the captain's; and when topics of interest failed, owing to the paucity of my comprehension, we resorted to fingering each others' ornaments, and exclaiming 'Bagus, bagus' (pretty, pretty), like a mutual admiration society.

Breakfast, tiffin, and tea were rolled into one at twelve o'clock, when I sat down to a repast with the captain and the mates. The poor captain had great difficulty to find time to swallow, so much his duty did he consider it never to interrupt his own flow of words. My presence caused the first mate to be covered with confusion and blushes. In his embarrassment he alternately fed himself and the ship's cat off his fork, and then dug it into the contents of the dishes from which I was supposed to be helped.

I slumbered most of the afternoon, waking at intervals to answer the captain at random, and going asleep again to dream that the Chinese passengers were rising to murder us. There was no earthly reason why they should not have done so had they so desired, except that the Chinese are a law-abiding people and do not rise in that promiscuous way. I had a comfortable conviction that they would not, and so was able to enjoy a little superficial thrill of fear that they might....

At last the captain woke me up by demanding my passage-money. I happened to have made sure of the amount before starting, so knew that he was mentioning an entirely fancy price. I did not think I needed to pay extra for the exercise his tongue had indulged in, so I told him agreeably that I could not think of troubling the commander of the steamer with petty finances, and should be obliged if he would send me the compradore. The captain looked as if he regretted his lofty position, and reluctantly called for the Celestial, who asked for the modest sum which was strictly legitimate, and made Chinese integrity on that occasion compare rather favourably with Scotch.

The vessel took nearly an hour coquetting with the wharf at Port Dickson before she could be induced to lie alongside. The captain told me I was lucky; she usually took an hour and a half....

I went off to make the most I could of the sea. There was no need to shiver on the brink before entering it, or be nervous of a cold shock. The bath had been carefully heated, the waves all warmed deliciously, the sea, like the atmosphere, tended to luxury; even the tide seemed to be filled with a lazy languor as it came creeping up over the beach.

Ethel Douglas Hume, *The Globular Jottings of Griselda*, William Blackwood & Sons, Edinburgh, 1907, pp. 91–3.

11
The Triang River in Flood

ARTHUR KEYSER

In the mid-1890s Keyser was District Officer of the remote Jelebu district of Negri Sembilan and he had to cope with one of those sudden spates to which Malayan rivers are subject. Passage 27 below deals with the more widely known Kuala Lumpur floods of December 1926.

I have a less pleasant incident to record in connection with the Triang. Unprecedented floods had converted valleys into lakes, and no one could tell where they ended or rivers began. Two Malay police were brought to me by the sergeant-major with the story that they had been several days coming from the down-river station, resting at night in branches of trees. This tale, although sounding improbable, turned out to be true. When interviewed, these travellers told me of villages isolated by floods, houses washed away, families living in trees and no boats for rescue. I called for volunteers, and two dug-outs were at once loaded with rice and flour and started off to take relief. I could not sleep that night for thinking of the occupants of those boats who had so little chance of reaching their destination. In the morning, therefore, taking two boats and selecting a couple of experts in this river navigation, I started down-stream. It was a constant difficulty to avoid swamping by collision with fallen trees, dead buffaloes, pigs, and other victims of the flood, which the roaring current was sweeping along with our boats. Those generally fearsome rapids no longer existed, as we skimmed lightly over rocks submerged many feet below. An island where we had camped with the merry crowd of tuba fishers, together with our frail buildings, had long preceded us down-stream. But owing to good fortune and the wonderful skill of Awang and Soho, my experienced boatmen, we eventually reached our destination without mishap, relieving one small village on the way. The boats of yesterday were also safe, but only arrived one hour before ourselves.

Then commenced the work of picking people off the roofs of scattered houses and taking them to the police station, which was of stouter build, where also on the roof, they were fed. Ultimately all were conveyed to the safety of a neighbouring hill. Our return journey on foot, well inland, proved rather arduous, for it was dizzy work wading long distances through rushing water and occasionally falling headlong into deep places, spanned at normal times by open culverts or bridges.

On arrival at my little headquarters we received a great welcome, as many had prophesied that we would never return.

A. L. Keyser, *People and Places: A Life in Five Continents*, John Murray, London, 1922, pp. 150–1.

12
The Dam

ARNOT ROBERTSON

The origin of the attempt to walk overland through part of Trengganu has been described above (see Passage 2). The party run into a number of obstacles, in particular a river which bars their direct route. This passage describes an attempt to cross it which almost ends in disaster.

THEN the twisting river bent westward, and we were forced farther into the interior.... Once we made an attempt to get across the river, when the continual westward trend of our progress forced us to realize that we were wasting our time and our little remaining energy. We were covering shorter distances every day, through one thing and another. There was no heart left in any of us, but we went on because we had no option.

We found that floating debris had made a loose dam across a narrow neck of the river, below which the stream broadened again, running turbulently among half-submerged rocks. It was the only feasible spot we had struck, and we were readier to take risks than before, though less able to deal with them. It was hard to reach the actual bank here, protected as it was by swamp, and it took us a long time to wade through to the edge of the dam.

Several heavy tree-trunks, their roots and branches intertwined and holding for the moment, were as far as we could discover the only anchorage of the constantly thickening jam of odd material.... Upstream the current ran rapidly through a little gorge. We could not cross by that side; we should be sucked under the dam and inextricably tangled in the mass of roots and branches protruding under water. On the downstream side, where the water spread out round rocky shoals, the current could probably be stemmed with the help of a handhold on the dam; but we had no means of gauging the security of the interlocked flotsam. The whole thing looked appallingly insecure to me, and if it broke up while we were making our way across we had a thin chance of surviving the rocks....

The water sucked viciously at our legs before we had waded half a dozen steps through the shallows. As we lowered ourselves farther into it our bodies grew light and helpless and stones moved under our feet as we strove to find some hold in the river-bed, against the rush of the stream. No one could look for help from the others; it was a full-time job to get oneself across. Deotlan stowed the little ammunition we still had in his headcloth, and gave his rifle to Arnold and Stewart to look after between them; they were both a good deal taller and had a slightly better chance of keeping it dry. Arnold's revolver had been thrown away long ago, when the ammunition ran out: we should get no more of that at any rate.

I went third, with Deotlan following, but I had no knowledge of the progress of the others after the first minute. With one arm over half a log, I slipped on the unstable bottom as I reached out with my other hand for a branch farther along. Instantly, as though the water were a live thing waiting to pounce, a shoot of current swirled round me, sweeping my feet from under me. The soft, terrifying hands of the river tore at my shoulders. I felt the log shift slightly, not giving but rolling over towards me, so that my hold was weakened. I tore with my nails at the smooth bark as it turned, and found a crevice, but now it was only the crook of my fingers that held me against the suction of the river, and I was swinging a little in the changing eddies that formed through the thickening dam, as a rush swings in a half-choked brook. The pressure of the water was considerable and I knew that I should lose this grip if I put too much strain upon it, so that the log turned over completely. Soon my bent fingers began to tire. Frantically groping round, my free hand came on a submerged stick, caught in the lower side of the dam, but not securely, it felt. Gently I pulled on it, and it gave at once. I let go helplessly. I tried to shift hands on the log and could not; the crevice was too small. As a last resource I felt below me in the water again, for the loose stick, meaning to free it entirely and make an unlikely attempt to wedge it between the log and whatever was on the other side to act as a lever. The stick came readily for some way, and then held when I thought I had freed it. I could not tell how safely it was caught before I let go of the log as it suddenly turned farther. I drew myself half under water but close against the dam by the stick, and, kicking against the yielding water, made a wild effort to reach a thin root or piece of broken creeper that stuck up out of the dam above my head. My left hand touched it, bent it, and slipped along it, but I managed to hold on, near the end. It gave threateningly under my weight, but did not break. Letting go immediately with the other hand, I was swung round by the water to a position in which I could reach a rough-barked tree, and pull myself head and shoulders clear of the water for a breather....

As I worked my way out into midstream, the tug of the water increased beyond my expectation: but it may have been that I was losing courage and it merely seemed to strengthen unduly fast. I was becoming panicked by the relentlessness of the current, as one is almost unnerved sometimes when sailing by the merciless persistence of a high wind. If it would let up for the space of a breath, vouchsafing even a second's rest, it would not be so terrible; but it goes on without a second's intermission. Now, at each shift of my hands, water tried to slip between them and their objective, water-cushioned every outward swing, deadened my efforts, so that, gasping and afraid, I reached short of my aim, and at each small error in judgment, water poured chuckling over my shoulders, sometimes over my face.

Always uncertain of the strength of my hold, I progressed only a foot or less at each anxious move, and occasionally not at all for long seconds at a time, when everything I touched felt too yielding to trust. Eventually I would be forced to depend on it, for want of better support, and by luck it always held. There was one particularly bad patch, where a sapling was caught only by its branches; the slender trunk, floated away from the dam by the stream, was very nasty to work round: it cost me a struggle to let go of the dam and rely entirely upon its flimsy hold, yet it was impossible, for me at least, to duck under the trunk in this fierce current: I tried twice and failed. Long parasitic creepers clung to the young tree, and I felt the water gently folding these still-living streamers about my head and shoulders and arms, as I tried to force myself down under water. I had no idea how the two who went in front of me had negotiated this bit....

[As they came back to rescue me] trunks turned under their feet, small branches broke, and they jumped wildly for another patch of tangled rushes and creepers and grass, and leapt on before that had time to give way entirely. One of them—I did not see which—straddled two logs and leant out perilously to reach me, while the other lay full length on a light mesh of rubbish, that should not have held for a second. I was hauled up regardless of the ripping edges of broken branches and saw-like grass: I did not realize till later, when I began wondering where all the gore in my vicinity was coming from, how much they hurt me between them while getting me out.

Stewart, well past midstream, had been making his way along a difficult stretch towards the goal of a fair-sized log, a few yards ahead, when the log stirred, curved lazily in a half-circle, and sank. Some drowned and partially eaten babirusa [wild boar] were caught in that side of the dam, nearer the bank. He saw several spiny tails, that he had taken for boughs, flip up gently and disappear as he yelled and scrambled unavailingly among the breaking, yielding debris, threshing water with his feet as he tried to find a solid hold. It took him what he described as a long while to climb out: but it could not have been more than a few seconds of mental agony.

All this, naturally, he and Arnold did not tell me at the time, but one Malay word '*Buaya!*' [crocodile] shouted by Arnold to Deotlan was enough to make me understand. It was enough for him too. A fisherman, he was more agile in these surroundings than we were; he was half out of the crumbling dam before Arnold yelled to him again.

We were lucky that the beasts were gorged: but a crocodile

kills wantonly, for the sake of killing: it would be as easy for them as for us to get from the water on to the dam. We could not stay where we were, and the strained, haphazard structure on which we crouched would not bear us if we tried to crawl along it. But we were relieved of the need to make a decision on which everything depended: for once Malaya played upon us a more or less kindly practical joke—detestable, like all practical jokes, but ranking as kindly because it was not the worst that might have happened....

The dam gave, near the other end. That was a ghastly sample of eternity, the endless stretch of time in which we felt the thing buckle and slip, jerking itself free as the stream tore, one after another, the last strands that anchored it. At the bank from which we had come it held for a few instants longer, and those instants saved us. Like a gate opening, the remains of the dam swung round downstream, disintegrating as it did so. The part where we clung reached the shallows before it broke up altogether, depositing us within easy reach of the bank from which we had started.

E. Arnot Robertson, *Four Frightened People*, Jonathan Cape, London, 1931, pp. 177–84.

13
A Visit to Trengganu and Kelantan in 1846

ANON

This passage is from an anonymous article published in the Journal of the Indian Archipelago. *There is reason, however, to believe that it was written by J. D. Vaughan (the author of Passages 49, 68, 75, and 86 below). It describes a voyage of the East India Company's patrol ship* Phlegethon, *to Trengganu, where the Raja (Sultan) was Baginda Omar (r. 1839–76), and then to Kelantan where the Raja was Sultan Mohamed II (r. 1839–86). These were among the ablest Malay Rulers of their time.*

Captain Ross of the Phlegethon *was sent to secure the*

release of a quantity of tin, the property of Singapore Chinese merchants, detained by the Sultan of Kelantan. Passages not reproduced here describe the difficult, but ultimately successful, negotiations leading to the handover of the tin. (On Vaughan's varied career, see Passage 49.)

ON the 3rd April, the steamer came to an anchor off the entrance of the Tringanu river, and was boarded by a police boat. Captain R. intimated his intention of visiting the Rajah early the next morning and desired them to have fire-wood ready for us. On the 4th, the visit was paid and a letter or message from the Governor of the Straits Settlements, Colonel Butterworth, C. B., was delivered. The town or village of Tringanu has a very pretty appearance from the river; the houses line the south bank for a couple of miles and are built very close, on piles; in the back ground are two hillocks, on one of which the standard of Tringanu is hoisted on great occasions, and behind the hill is a forest of cocoanut trees; on the north bank there are a few isolated houses, built in groves of cocoanuts. Some of us were employed loading the ship's cutters with fire-wood some dis-

A typical riverside village, with its mosque in the centre, from Major Fred McNair, *Perak and the Malays: 'Sarong and Keris'*, Tinsley Brothers, London, 1878.

tance beyond the town, and while doing so a very respectable Chinese invited us to his house, where a slight repast was served up and our host surprised us by producing some capital French brandy and enjoyed his glass as well as any of us. In the afternoon the Captain received a present of bullocks, fruits and cakes, and the Rajah sent a message that he was desirous of visiting the steamer.

Early on the 5th he came off in great pomp to his schooner the 'Dragon' and there waited till our boats were sent; he soon came on board, and in a few minutes the decks were crowded with his followers. The Rajah is a fine looking man, very plainly dressed in a sarong, white baju and black velvet jacket slashed with gold thread, very shabby and worn out. His body guard was composed of 12 Malays, armed with krisses, and each man carried a drawn sword in his right hand; they looked a filthy set of rascals—in dress they do not differ from their rabble. The heir apparent is a fine lad, about 4 years old, and was carried about in the arms of a follower, who kept close by the side of the Raja. After minutely examining the engines and other parts of the vessel, the Rajah, his Prime Minister and nobles were treated in the cabin, whilst the mob on deck was amused by our band, consisting of a drum and fife. Some of the natives had learnt the art of ship-building at Singapore and assisted by Chinese carpenters had built the schooner abovementioned, which was a very creditable piece of workmanship. She is rigged well and kept exceedingly clean. Wood, therefore, they had seen worked into ship-shape, but how our ship had been built of iron they knew not. It was strange to see them examining the sides, knocking the iron, and looking at each other quite puzzled.

A very curious incident occurred below, which amused us not a little. The Rajah took his son in his arms, caressed him, fondly patted him on the cheeks and then suddenly gave him a sharp slap on his nose, which brought tears to the boy's eyes. The little fellow looked up and smiled as if quite accustomed to such tokens of affection. I understood from one of the guard that it was to flatten the nose, the boy having rather a prominent one, which is not considered a mark of beauty among the Malays.

In a couple of hours all had quitted the vessel and after taking in sufficient wood, we steamed for Kalantan. The scenery changes gradually to the northward, the hills are much larger and some distance from the coast a very high range of mountains is seen. We anchored in the evening off Kalantan and fired a gun to attract the attention of the people on shore. Shortly afterwards a guard boat came off, the steersman seemed very much alarmed, he was seen waving a white cloth over his head, and trembled so much on gaining the deck he could scarcely speak. This man was well known to the Chinese passengers, who stated that he was a great tyrant, extorting large sums from the native traders, and is the chief of the village at the mouth of the river. The strange appearance of the steamer and the gun we fired, no doubt disturbed the poor man's conscience.

On the 7th Captain R., accompanied by 2 officers and 20 men, well armed, ascended the river to negotiate with the Rajah....

Early on the 9th we went up once more, when the Rajah Kechil met us and said his brother was not prepared to receive us. He had us conducted to his own house, where his wife received us, and presented us with fruit, seree* and sweetmeats. She is a very fine woman, between a Chinese and Malay, handsomer than any native woman I had seen in the Straits, and very lady-like in her manners. She is the mother of the Chinese Captain. Her late husband, a Chinaman, who died some months ago, was Captain for many years and at his death the Rajah Kechil seized her and took her to wife by force. This displeased her very much, not that she took any objection to the man but that it deprived her of the influence over the Chinese, about 1,000 men, which she acquired on the death of her husband, her child being too young to undertake the duties of the office. The Chinese were equally enraged and expressed themselves very strongly on the subject when not in the presence of their ruler, but said they were obliged to submit.

The house we were received in is surrounded by two wooden enclosures about 18 feet high and 100 yards apart; the house has the appearance of 3 huts joined length-ways, with no partitions dividing them, or one house covered by a 3 peaked roof; it is about 150 feet by 100 feet. From the entrance the floor rises by steps 8 or 10 feet wide to the

*The offer of betel-nut (*sirih*) and other ingredients for chewing was a traditional Malay courtesy to welcome a visitor.

musnud, dais or platform on which the Rajah sits. At their meetings the natives of the highest rank sit near the chief, the next grade on a lower step and so on, decreasing till the common people sit on the ground. Behind the musnud is a door leading to the female apartments or harem. During our interview a great number of females were seen peeping at us, but the screens prevented our criticising their beauty. From the roof hung howdahs, seats and trappings used on elephants. We saw several of these animals feeding in the yard....

On leaving the Rajah, the little Chinese Captain met us and begged we would partake of a dinner prior to our journey down the river. To this we assented and were very hospitably entertained by himself and followers. The lad is about 14 years of age, of an agreeable appearance, of a slight frame, marked by small pox, and partaking largely of his mother's manners. Three Chinese sat with us at dinner, and a crowd of inferiors sat in the room and looked on with curiosity. The Chinese are excellent cooks, and will convert the most simple materials into a very savoury dish. The ceremony observed was this;—before a morsel was eaten, tin cups, holding little more than a thimble full of arrack or samsoo, were placed before each individual and at a signal from one of the Chinese, all had to raise the cups to their lips, empty them and place them on the table again. The chop sticks are then seized and held in the air; at a signal, every man falls to and eats as much as he can. This was no easy matter, my companions fared better than myself, having learnt the use of the sticks in Cochin China [Vietnam], but for the life of me I could not pick up a single piece of meat. Our host finding me so ignorant, ordered in an English fork and spoon with which I made up for lost time, much to the amusement of the spectators. After a few mouthfulls the master of ceremonies stopped, dropped the sticks and raised the cup. We were obliged to follow the example, hobnobbed and returned to the sticks, this was repeated several times. Chop sticks are two round pieces of wood, bone or ivory, about 10 inches in length, they are cleverly held between the fingers of the right hand and the food seized by the tips. This can easily be done with a little practice as the food is always cut into small pieces. I have noticed however, that when very hungry they lift the dish close to the mouth and shovel in the food with both sticks held close to each other.

When about to return R. received a very handsome kris, the handle being inlaid with gold. On the 10th, all the articles detained were sent off as promised, and we returned to Tringanu, where we were entertained by the Rajah, and on the 11th the Rajah's uncle with his followers came on board. One of them, to my surprise, instead of saluting us on coming on board with the usual 'Tabeh Tuan', said in excellent English 'Good morning, Sir'. Concluding that he had picked up these words in his intercourse with Europeans, I was walking away when I thought he said 'I am a Waterloo man'. This attracted my attention and on questioning him he related the following particulars of his eventful life: He is a German named Martin Perrot, joined Napoleon's army as a conscript when very young, served in Italy and Germany and subsequently in Spain and Portugal, was taken a prisoner by the British, entered their service, fought at Waterloo in the second Brigade of German Rifles and received the medal for that victory, received his discharge on Peace being concluded, and came out to India in the Dutch service. On Malacca being abandoned he left and settled in Tringanu, adopted the native costume and turned Mahomedan, carries on his original trade of blacksmith, by which he obtains a comfortable living, declares he has a perfect recollection of Wellington, Napoleon, and other remarkable characters. At first sight I took him for an Albino, several of whom I have seen in the Straits, and in his native attire, reddish brown skin, blue eyes and white hair, he certainly had more the appearance of one than of a European. He has a perfect command of our language, and no doubt has been a soldier from his knowledge of military tactics, but I much doubted the story of Waterloo etc.

Anon, 'Journal kept on board a Cruiser in the Indian Archipelago in 1846', *Journal of the Indian Archipelago*, Vol. 8, 1854, pp. 175–80.

14
Piracy at Kuala Langat in 1873

MAT SYED

In the 1870s small sailing boats (in this case a nadir *or 'naddy') still carried much of the local Malayan trade and passenger traffic. A crew of half a dozen was headed by a master* (nacodah) *and boatswain* (juragan). *To deter robbers they carried muskets, spears, etc. but the attackers were adept at ruses by which they could approach their target until they were close enough to make a sudden, overpowering assault.*

The episode described here is famous as the pretext for British intervention (and assumption of control) in Selangor in February 1874. The Sultan (Yam Tuan) had little authority and it is clear that one of his sons (Tunku Allang), and his followers, were implicated. The trial was conducted by a mixed court of Europeans (the president being a Singapore lawyer, J. G. Davidson, whose notes of evidence formed the record) and Malay notables.

Many years later (in 1900), after Davidson's death, Swettenham, who did not come to live at Kuala Langat (as the Sultan's adviser) until several months after the trial, asserted that (in 1875) he had obtained conclusive evidence (from unnamed informants) that the accused, who were convicted and executed on a charge of piracy, were innocent, since none of them was present at Kuala Langat when the attack took place. The accused (here identified by numbers) put questions to Mat Syed, the key witness, to challenge his identification of them, but none asserted that he himself had been elsewhere at the time.

MAT SYED, sworn, states—I live at Tranquerra, in Malacca, and am a seafaring man; I left Langat on 25th of the month of Poasah* in a naddy belonging to Malacca; there were three Chinese passengers, whose names I do not know, and six Malays belonging to the boat named Hadjee Doraman, who was the nacodah, Ah Kim, Tamb Itam, Meman, Mambi, and myself. The naddy was

*Puasa; Muslim month of fasting.

loaded with rattans; there were also boxes. There were 2,000 dollars on board belonging to Ah Kim, of Langat, and the nacodah. I assisted to bring the dollars on board the boat and the nacodah told me there were 2,000 dollars. We left Bandar Langat about 6 A.M.; we arrived here (the stockade at the Qualla [estuary]) about 1 o'clock, and showed our pass to Arsat, who was in charge of the stockade. We went outside the river about a mile and anchored because the wind was against us. We anchored about 3 o'clock; the nacodah told us to rest, and we would sail at night. About 3 o'clock, the juragan called the crew to boil rice. We cooked rice, and about 5 o'clock I saw two boats coming out of this river. I asked the juragan what boats they were, and he said two friendly boats from the stockade. They pulled up near us, and Doraman called 'apa kabar', and the boats replied 'kabar baik'. Doraman asked where they were going, and the reply was, they were going to fish. Musa (No. 4) replied from the boats. One of the boats came alongside, and Musa and three or four others came on board. The other boat came alongside on the other side. (There were about twenty men on the two boats.) They talked to Doraman. About 6 o'clock Doraman told us to bring the rice. When he was to begin eating shots were fired from both boats. Doraman fell to the shots. Musa then called out to 'amok'. Three of our people jumped into the water and were stabbed, and all the others in my boat were also stabbed and killed. I jumped into the water, hung on to the rudder, and after dark floated away to the shore; when I floated away the three boats were still together in the same place. I floated to the piles of this jetty and got hold of one. There was a Bugis boat lying about three fathoms off. I held on to the pile about an hour, and the pirates came in their own boats bringing Doraman's boat with them. One man came out from the stockade on to the jetty and asked 'Sudah habis'. From the boats a man replied 'sudah habis. We are taking the property to Tunku Allang.' They all went up the river with their boats and my boat. The two boats returned in about an hour without mine. They all came up on this jetty. People from the stockade asked if it was finished, and they said it was all finished. After all was quiet here, I went to the Bugis boat and asked them to assist me, and they took me into their boat. The Bugis asked me whose boat it was, and I told them Doraman's. The Bugis advised me not to say anything about the affair here or I would be killed. When I was speaking to the Bugis the people from the stockade came and asked them for me. The Bugis refused to give me up, but said they would show me next morning. All the prisoners were in the boats that attacked us. It was daylight and I could see them quite well. No. 1 shot the juragan. No. 2 came into the boat and shot and stabbed people. No. 3 remained in his own boat and had a spear. No. 4 came into our boat and stabbed Tambi Etam. No. 5 was in his own boat, he had a spear and stabbed people in the water. No. 6 came into our boat and stabbed Meman and others. No. 7 remained in his own boat and had a spear and assisted to stab my friends in the water. No. 8 was in one of the boats, and I did not see him do anything.

The next day the Bugis took me on the jetty and showed me to the headman of the stockade named Marsat. All the prisoners were present with Marsat at the time. Nos. 1 and 4 asked the Bugis to give me up, but the Bugis refused. I saw on the jetty the boxes of many of my friends, also bags belonging to the boat scattered about, and met also two muskets belonging to my boat and one spear and a sword. (Marsat is produced and witness identifies him.) The Bugis then took me up to Langat to the Yam Tuan, who was asleep, and then they took me to the Datu Bandar. He asked me if I knew the men who had done it. I said I did. He then asked me where they belonged to, and I said to the stockade. He then said to me if you are asked say you do not know who had done it, if you say you know them you will be killed. After this the Bugis took me back to the Yam Tuan, and I told him all that had occurred. He then asked me if I knew the people who had done it, and I said no, as I was afraid of being killed. When I was speaking to the Yam Tuan, No. 1 and No. 4 came in and said we want this man (pointing to me), Tunku Allang wants to take me to the Qualla. Yam Tuan told me to go with them. I said I was afraid. Yam Tuan said if I was afraid they had better let me go to Mahomed Syed's shop. Mahomed Syed was present and I was given up to him. Mahomed Syed sent a letter by Belal Ismain to my uncle Mamoot, at Malacca. My uncle arrived at Langat from Malacca on the 27th of the month of Poasah, and the same day I left

Langat with my uncle. When I was leaving I saw our boat at Qualla Sungie Durien in this river; no one was on board. Qualla Sungei Durien is Unku Allang's place.

About twenty days after my arrival in Malacca, I was on the bridge at Malacca, and saw two boats coming up the river, and saw some of the prisoners in the boats; I reported to Mr Hayward, and I went with Duffadar Mahomed and pointed out Nos. 1, 2, and 3, and Mahomed arrested them in their boats.... In the boat where Nos. 1, 2, and 3 were arrested we found a musket, a sword, and a spear, which belonged to Doraman, and were in his boat when she was plundered....

Two days ago I went up the Langat river in one of the man-of-war's boat, and saw Doraman's boat inside the Sungie Durien. It was tied to the mangroves as if hidden. There was a house on shore near the place; it belongs to Tunku Allang. The naddy was then brought down the river, and is here now. This is the naddy. After coming down with the naddy, I came on shore here, and found a water cask in this stockade belonging to the naddy, and which was on board when it was plundered.

Evidence of Mat Syed, part of the record of the trial enclosed with Straits Settlements despatch to the Colonial Office, dated 24 February 1874.

15
The Defence of the 'Lizzie Webber'

JOHN DILLON ROSS

Like the preceding passage this story of a pirate attack is a famous episode, which occurred in July 1862 off Labuan. The author was the four-year-old 'Little Johnnie' of this story of his father's eight-hour defence of his ship.

The Illanun (and Balanini) pirates, the scourge of Malayan waters in their time, had their bases in the southern Philippines, from which they sailed forth—usually to attack comparatively defenceless small native vessels. The Lizzie Webber *became their target because her cargo included 'a large sum of money'. The story brings out the tactical advantage, in sheltered waters, of oared galleys against sailing vessels and the reluctance of pirate captains to incur heavy casualties. Persistent efforts of British, Dutch, and Spanish warships, including steamships which could overtake and outmanoeuvre galleys, eventually destroyed the pirates and their bases.*

THE brig sailed one fine afternoon, with a light and fitful breeze. During the night the current set her over towards the Brunei coast. At daylight, as she slowly rounded a point, her captain saw, to his horror, a fleet of Illanun vessels lying in wait for him in a little bay. The wind died away altogether and left the brig without so much as steerage-way on her.

All was hurry and bustle on board the *Lizzie Webber.* The men beat to quarters, the guns were run out, and rifles and cutlasses served to the crew. Captain Northwood went below for a few moments, during which he handed his wife a revolver, with strict injunctions to keep in her cabin during the fight, and to shoot her son and herself if the pirates carried the ship. This son was little John Dillon Northwood, born in Singapore some four years before.

Northwood took up his post on the quarter-deck while events developed rapidly. The pirate squadron came sweeping down on their prey, each prahu [vessel] pulling some forty oars or more at a great rate. The pirates were several hundreds strong, and had they gone straight to work, the fate of the *Lizzie Webber* would have been settled within the next half hour. That, fortunately, is not the native way of doing things. The pirate vessels pulled right round the motionless brig and then the leading prahu swept towards her until both vessels were within easy hail of each other. An exposed platform had been erected right amidships of the pirate vessel, on which stood, conspicuous in a scarlet jacket, Si Rahman himself. The pirate chief hailed Northwood by name and explained that he had come as a friend and, being short of tobacco, proposed to come on board the *Lizzie Webber* to purchase a supply of it. To this Northwood's reply was that he knew pirates when he saw them, and that if the vessels which surrounded him did not sheer off, he would open fire on them at once. The Illanun adjured Northwood to give up his ship without a useless struggle, especially as he himself wore a magic charm which rendered him invulnerable to shot or steel.

Cassim, on the main-deck, was anxiously watching his master's movements, and thinking that a certain wave of the hand gave him liberty to do so, immediately fired the gun of which he was in charge at the pirate prahu, and thus commenced the action on his own responsibility, much to the anger of Mr Simpson, the chief officer.

The *Lizzie Webber* carried a battery of six twelve-pounders, and Northwood, having purchased some time previously a few cases of American muskets from a Yankee captain whom he had met at sea, had half a dozen muskets for every sailor on board. These had all been carefully loaded before the fighting commenced. Thus the *Lizzie Webber*'s crew had only to throw down an empty musket to pick up a loaded one, and kept up a rapid fire which greatly disconcerted the pirates. The brig, like other ships on that coast, carried a very large crew to work her cargo as well as to sail her. Her twelve-pounders were each loaded with a round shot and a canvas bag of bullets rammed home on top of it, which made a very effective charge at short ranges.

From the moment that Cassim fired the first shot, the *Lizzie Webber* kept up a hot fire with her guns and musketry, while the pirates, who mounted a number of light guns, replied vigorously. Their prahus were strengthened by breastworks of Borneo ironwood strong enough to withstand the impact of a round shot.

The roar of the guns and the constant rattle of the musketry made a terrific din, above which rose the yell of the pirates. Mr Simpson was carried below badly wounded, with three native sailors in the same condition, within a few minutes after the action commenced. In the absence of any surgeon, Mrs Northwood had to attend as best she could to the wounded. Poor lady! she had trouble enough on her hands. Her son Johnnie, so far from being frightened by the din of battle, made frantic attempts to escape to the upper deck to see what 'papa was doing'. He had finally to be carried off by his mother, kicking and screaming with rage, to be locked up in a spare cabin.

Mrs Northwood's presence of mind saved a disaster, which would have put any further resistance out of the question. The *Lizzie Webber*'s magazine was right aft, and was got at through a scuttle in a store-room. As powder was running short for the guns, Mr Jenkins, the second mate, sent a couple of native sailors to bring up some kegs from the magazine. With the usual thoughtlessness of Malays, the sailors were actually going into the magazine with a naked light, when Mrs Northwood rushed in time to seize the flaring lamp and throw it through an open port-hole into the sea. She then went into the magazine herself to see the powder handed up, and found the whole floor of the place covered thickly with loose gunpowder. In those stirring days a captain's wife needed to have her wits about her!

In the meantime Captain Northwood himself was fighting his ship for all he was worth. His great object was to bring down Si Rahman, who exhibited the most extraordinary daring—possessed as he was of the idea that his magic charm would preserve him from all danger. There he stood on his platform, like a scarlet demon, directing the attack, and constantly exposed to a rattling fire, and somehow nothing could touch him....

From the main-deck and forecastle scores of shots were directed at Si Rahman without hurting. The man was perfectly aware of the unavailing attempts to bring him down, and openly rejoiced in the strength of his magic charm. It really seemed as if the charm was working to some purpose. Cassim, especially, tried the united effects of round shot and bags of bullets on his hated enemy, but while they took full effect on the crew of the prahu, nothing could touch the scarlet Si Rahman.

After three hours of desperate struggle there came a lull. At a signal from Si Rahman all firing from the pirate squadron ceased, and they pulled away from the *Lizzie Webber*. For a brief moment, Captain Northwood hoped that they had had enough of it, and that the attack was definitely repulsed. Nevertheless, he went round his decks, saw that the heated twelve-pounders were sponged out and ready for further service, and that all the small arms were reloaded. Barely had he completed his round when he saw the pirate vessels sweeping down upon him again under the full pressure of their swift oars. Si Rahman's plan of battle was now evident. Having at first hoped to make an easy prey of the *Lizzie Webber*, when he found himself baulked by the deadly firing from the brig, he tried to wear down Northwood's resistance by picking off his men, and exhausting his ammunition. Now, disappointed with the results of his tactics, he decided to do what he should have done some hours before, and, throwing his hundreds of men on the deck of the *Lizzie Webber*, leave cold steel to do the rest. The eight prahus made a dash for the starboard side of the brig, thus leaving her port battery idle.

Northwood saw the coming onrush with the blackest despair. How could he hope with his scanty crew to withstand the onslaught of some hundreds of desperate Illanun pirates? He had half a mind to throw up the sponge, to rush below, and after despatching his wife and child to blow up the magazine. However, the fighting instinct is hard to quell in such a man, and for another minute he watched the occasional shots from his starboard guns flying harmlessly over the approaching prahus, which lay so low in the water that it was impossible to depress the muzzles of the *Lizzie Webber*'s guns sufficiently to hit them. Si Rahman's own prahu, which was by far the best manned of the pirate fleet, drew rapidly ahead of the rest, and was almost alongside the brig when Northwood saw Cassim about to fire his twelve-pounder, with a result which must necessarily be harmless. Leaping down to the main-deck, Northwood restrained the Malay's hand and shouted: 'Don't fire over the prahu! Train the gun straight on the platform and kill Sidi Rahman!' Cassim showed that he had driven the wedge beneath the gun's breech as far as it would go and that he could depress the gun's muzzle no further. Looking round with a hunted desperation in his eyes, Northwood happened to see a spare spar lying on the deck. Suddenly bending down he put forth the whole of his great strength, and lifting the gun-carriage bodily, he got Cassim and another sailor to roll the spar beneath it. Then taking a hasty look along the sights he fired the gun. Before the smoke had cleared away, dire yells arose from the pirate prahu, while shouts of joy rang from the decks of the *Lizzie Webber*. What Captain Northwood saw when he looked over his bulwarks was a pile of wreckage in place of the famous platform on which Sidi Rahman had so recklessly displayed his scarlet coat. As for Sidi Rahman himself, a scarlet patch in the water swirling around the sinking prahu sufficiently accounted for the fiery Illanun chief. His magic charm had failed him at the critical moment.

The other pirate vessels rescued as many as they could from the sinking boat, under a hot fire from the *Lizzie Webber*, and then pulled away from her. And now, a gentle breeze at last springing up, the *Lizzie Webber* began to get way on her. Captain Northwood thought once more that he had escaped from the toils of his enemies. Still he availed himself of this respite to put his ship in fighting trim. Guns and small arms were all loaded, and the tired crew given their first meal during this terrible day.

But the pirate fleet still hung round the brig, their boats could still pull much faster than she could sail, and it was clear that the foe was only waiting for darkness to deliver his final attack. It was a grim and hopeless prospect. However, the breeze was freshening all the time, and as the *Lizzie Webber* began to slip through the water, the pirates, thirsting for revenge as well as plunder, closed on her under the rays of the evening sun.

Once more the firing raged hotly on both sides, when suddenly in the midst of the fighting Mrs Northwood appeared on the quarter-deck to tell her husband that there were only six more kegs of powder left in the magazine. It was not a message she could trust to other lips than her own. 'Very well!' said Northwood. 'Send up the six kegs and leave the rest to me. Obey my orders and go to your cabin.'

The pirates were rapidly approaching the *Lizzie Webber*. The six kegs of gunpowder would just enable her guns to fire one round apiece, with something to spare for the muskets.... Suddenly he altered his tactics, and instead of flying from his pursuers, he attacked them. Down went his helm, he spilled his sails, and before the pirates could comprehend his new move, he was sailing through their fleet, firing his last broadside. One of the prahus was caught at a disadvantage. Springing to the wheel, Northwood altered the *Lizzie Webber*'s course by a few points. Next minute there was a tremendous crash as the keel of the brig rode over the wreck of the prahu. Some of the pirates were shot as they swam in the water. Others, with the usual agility of natives, actually managed to climb up the ship's chains in an endeavour to reach the deck, but were promptly cut down. Then came the sudden darkness of the tropics, and in its welcome obscurity the *Lizzie Webber*'s sails filled and bore her into safety.

John Dillon Ross, *Sixty Years' Life and Adventure in the Far East*, Hutchinson, London, 1911, pp. 26–33.

16
Amok

RICHARD SIDNEY

Richard Sidney came to Malaya in the mid-1920s to take up the appointment of headmaster of the Victoria Institution, Kuala Lumpur's premier secondary school (Passage 27). He was an imaginative, but not always successful schoolmaster (he later held other teaching posts in Malaya), but from the outset he showed a flair for journalism, writing articles for English provincial newspapers which he later collected and published in two books. He was for a time the editor of the Penang Straits Echo *(Passage 32), but that too did not prosper.*

In this passage Sidney tells the story of the most celebrated case of amok in this century, which took place on the SS Klang *on 31 October 1925, just after she had left Singapore for Malayan ports. It was fully reported in the newspapers but Sidney provides a thin veil of altered names and some reasonable supposition as to the causes of Hassan's insane violence; Hassan did not survive to be questioned. In the actual massacre, thirteen people, including the ship's captain, were killed and five more were seriously wounded.*

In its original sense amok (now spelt 'amuk' and in English, 'running amuck') denotes reckless, suicidal courage in battle. In a second, more sinister sense it came to be applied to mindless killing, often a release from a pent-up sense of shame, injustice, or frustration (the Malay term is 'sakit hati'—sick at heart). In the nineteenth century, cases of amok were common but, in different social conditions, it has now almost disappeared. Theories about amok, to which psychiatrists have made their contribution, are numerous—and often contradictory.

When the price of rubber was high, as it was in 1925, Malay landowners often received tempting offers (from neighbouring rubber estates wishing to expand) to buy their land. Hassan had come to Singapore to complete the sale of his land at a price which was to him a fortune.

HASSAN, though he had been up late, awaked early the next day, for he was anxious to conclude his business with a solicitor in Raffles Square. What a busy town it was and how fine were the shops, he thought. And the traffic....

The business did not take Hassan long, except that the solicitor had to send out to the Bank for the necessary notes, and his smiling rickshaw puller had waited outside all the time. 'Now to shop!' thought Hassan, and catching sight of John Little's windows he went in there, not without some foreboding, as he had never entered a European store before. He had promised his wife something English, and with all this money he could well afford....

Coming out he ran into a friend from up-country, and throwing his purchases into the waiting vehicle he came again on to the shaded pavement to chat. Hassan told his friend about the sale of the Rubber Estates, and could not conceal from him how proud he was that he and his family would now be able to move into a better house. The struggling days were over, henceforth his wife should work no more; even now he was bringing her a fine present. His friend was interested.

'When are you going back?' the other Malay asked.

'To-morrow morning by the train,' answered Hassan.

'Why not come with me on the boat this afternoon. Much cooler than the train, and we shall be there to-morrow morning. You can surprise your wife by arriving earlier.'

The idea appealed to Hassan; he would be able to see his family all the sooner—and that long, hot journey would be avoided. He agreed to meet his friend on the boat that afternoon, perhaps they could go aboard in a *sampan* [row-boat] together. His friend said he would arrange this, and they parted.

Bidding his puller hurry as he must get back to his lodging, Hassan lay back comfortably once more in his rickshaw; they passed innumerable shops—Stay! yes it was; a really fine *kris*, he *must* have it. He shouted to the Chinese to halt, and sprang down from the rickshaw and made his way into the shop. The obese old Chinaman wanted a great deal of money; Hassan grumbled, but paid; after all he was now a rich man, and he wanted to add to his collection. 'Be careful,' warned the Chinese, 'it's very sharp!' Unconcernedly the Malay placed it in the folds of his garments so that it was quite invisible. They went on....

Looking over the side the Captain saw two Malays coming up in a *sampan*; one was well dressed and had several parcels, and his friend helped him to put them on the boat; there seemed to be some understanding between the boatman and one of the Malays, for the *sampan* did not row away but remained near the ship. The launch carrying first-class passengers was now approaching, however, and the Captain's attention was diverted. For the next half-hour there was a considerable bustle with luggage coming on and passengers finding their cabins.... Further aft the sound of quarrelling can be heard, and if you had come to investigate you would have found that Hassan was not satisfied with his cabin. A Ship's Officer coming along tried to find out what all the trouble was, and Hassan explained that he and his friend wanted a cabin to themselves.

'Quite impossible,' replied the Officer, 'the ship is crowded; why even the first-class are full.'

For the first time Hassan seemed to lose control of himself, and onlookers saw a well-built Malay looking round with burning eyes. His companion tried to sooth him, and told him they must make the best of it. If Hassan liked he would go ashore and that would give him more room. Hassan said nothing, he went into his cabin. Presently he was out, and shouted for his companion who seemed to be making for the *sampan* which was still near the steamer.

'Where is that brown packet which had my wife's present?' he demanded furiously.

His companion denied all knowledge.

'You have stolen it!' cried Hassan, 'and you now want to leave the ship. I see your plan.'

The eyes of the Malay began to blaze with fury, and it was all that his companion and several onlookers could do to pacify him. By now the boat had begun to get under way, and reassured that his companion was definitely there Hassan moved to the rail and brooded on the loss of his precious packet. His wife's present! Gone! Would she believe him when he assured her that it had been stolen?... Hassan was stung by his thoughts, and went in again to his cabin to see if by any chance the packet was there. Nowhere! At that moment he remembered what he had bought for himself. Well he would pay out the thief at any rate....

'You've stolen my parcel!' shouted Hassan.

'I have not seen your parcel,' answered his friend.

And then without any warning Hassan drew from his clothes his long knife and plunged it straight into his companion's belly. The stricken Malay crumpled on to the deck. Shouts of horror arose on all sides, and Hassan staring wild-eyed found himself alone on the deck. Looking down he saw a pool of blood spreading, and hastily withdrew his right foot. What had he done? Killed him? He cared not. The dirty thief to take away the presents he was bringing his wife.... A face peered round the corner for a second—Hassan ceased his meditation and made for it. The rash man tried to get back into the shelter of his cabin but Hassan was too quick—he ran him through the back, and laughed as he saw a woman trying to conceal herself under the couch: lunging at her he noticed a small child run away and seek shelter; luckily he failed to injure it.... Two victims: he would teach them to try and rouse his wife's jealousy....

The panic was indescribable and was spreading all over the ship. The second-class passengers had bolted themselves in their cabins, and the deck passengers having nowhere else to go sought safety in the lower cabins of the first class and in the dining-saloon, in spite of the opposition of the Chinese boys. Hassan walking stealthily forward saw a deserted ship, with a head peering out here and there....

Meanwhile they had rushed up to the bridge.

'Malay running *amok*, Sir,' they told the Captain, 'has killed several already and he looks like coming up here.'...

As Captain Mac. descended to the main deck he was surprised to find every cabin door bolted, and to see only a Chinese boy. Aft he could hear screams and groans, and there were still some people rushing frenziedly forward.... At this moment Hassan came along holding his bloody *kris* aloft.

'You go round on my right,' ordered Captain Mac. 'We'll surround him.'

The Malay seeing the manoeuvre ran at the Captain who put his foot out to trip him, at the same time trying to seize his wrists: he was nearly successful for the Malay fell, and the Captain had hold of one wrist. But the hand holding the *kris* was still free, and as Captain Mac. attempted to seize the small man's other hand the dread knife was plunged into his abdomen. Just as the Officer who was to have taken Hassan in flank rounded the corner he heard Hassan shout for joy, and saw his old commander stumble away as if mortally wounded.

'Did he get you, sir?' he said rushing up to the Captain.

'Yes,' gasped the old man, 'take me to a chair, quick, and get me some brandy.'

Hassan now had the deck to himself, no one else venturing near him, but he noticed a launch approaching the boat, and heard shouts in Malay calling upon him to surrender. (They had got the Police quicker than they had hoped.)

Surrender! What use to surrender now? He had killed at least five; let them kill him if they could. He took up a strategic position so that if the police attempted to fire they would risk killing some of the people on board.... Unless he surrendered they would fire. Let them fire—he shouted back words of disdain. They would have to come aboard, and he could make that difficult....

Hassan heard a report; a burning pain seemed to pass through his left knee. He did not fall. So that was the best they could do was it. He laughed with derision at the puny firearms of the white man. Yes! he *would* go Forward; if he could only reach the dining-saloon there would be a fine harvest for his new *kris*...!

'Yes, that was a good shot. Necessary to fire to kill that time,' said the Police Officer as they bent over Hassan, now wounded in the chest and vomiting blood. His eyes still glared defiance, and those who ventured to approach him were afraid that he might get up and pursue them.

Upstairs they were carrying poor old Captain Mac. from the long chair to which he had been assisted after that awful knife had done its work. He had lived only a few minutes. After they laid him reverently down they felt a bulge in his right tunic pocket. It was his revolver: fully loaded!

R. J. H. Sidney, *In British Malaya To-day*, Hutchinson, London, 1927, pp. 85–92.

On the Roads

17
A Race for Freedom

AMBROSE RATHBORNE

Rathborne was an Australian surveyor who came to Malaya in 1880 and for the next fifteen years worked in partnership with Heslop Hill, in the business of planters and contractors. The firm played a leading part in the brief coffee boom of this period and in the even shorter European incursion into Selangor tin-mining in the early 1880s. As contractors, they undertook work on the construction of roads and railways.

In these activities Rathborne first worked in Sungei Ujong (Negri Sembilan), but later in Selangor and Perak. His activities brought him into contact with Malay villagers, whom the firm sometimes employed as labourers in their enterprises. His book is a reminiscence of his experiences, showing good observation and a sometimes quizzical eye for human foibles.

AN amusing episode occurred whilst travelling many years afterwards along this portion of the journey between Chanderiang and Kwala Dipang [in the Kinta valley of Perak]. I had driven to the former place, and wished to rejoin the cart road at the latter, and as over this intervening twelve miles a good bridle path had been constructed, the journey was much easier and less fatiguing than it used to be.

I had started to walk across the hills, and overtook on the way a Sikh policeman escorting a Malay prisoner, who was a noted housebreaker and thief. As we walked along my servants and the policeman entered into conversation; pedestrian feats formed the subject of their discourse, during which another Malay joined the party. The policeman became somewhat boastful of his prowess, and I could hear from the contemptuous way in which the Malays were answering him that his bragging disgusted them, and offended their sense of good manners and politeness. I was getting tired of the conversation, so quickened my pace in order to leave the man and his prisoner behind; not a bit of it, the policeman, after all his vain-glorious talk of how much better he could walk than anyone else, seemed unwilling to be distanced, and determinedly kept up. There were four miles of ascent in front of us until the gap at the top of the hill was reached, and as we tramped along I heard the prisoner expostulate and declare he could not travel at the pace we were going, but the policeman was too intent upon not being left behind to pay any attention, and, changing place with his prisoner, told him to catch hold of his belt and follow along, so as mile succeeded mile they both kept up, the Sikh striding on, and the prisoner running behind; at last the top of the hill was gained, and the descent began. Now came my opportunity; although the lithe and long-legged policeman was able to keep up with me during the ascent of the hill, I knew that I could distance him now, and made every effort to do so, for in front lay a long stretch of level ground without any shade, which I did not wish to traverse at a tearing pace in the middle of the day. I soon began to hear the clattering of feet behind, and knew that the policeman had also commenced to run, and was more than ever intent on not being left behind. Unfortunately for him, in his emulation he entirely forgot his prisoner, until a rustle was heard some distance back, caused by the prisoner making a dash into the jungle, and when we both turned round there was no one to be seen. The Malays had been distanced, and the prisoner had escaped. I laughed, as the policeman, instead of at once pursuing his prisoner, fumbled in his pouch to find a cartridge. He then fired in the air, and disappeared into the wood where the prisoner had vanished, and presently I heard a shot or two in the distance.

After a little he returned to beg me not to report him, as he was due for a pension very shortly, and started off at once again in search of his charge. In the meantime the Malays overtook me, and we completed the rest of the distance in comfort, and I saw no more of our sporting policeman. The affair was no concern of mine, and I was rather amused at the man whose unwillingness to be beaten had absorbed him so entirely and made him oblivious of his duty. Some time afterwards I made inquiries, and heard the prisoner had not been recaptured, and probably never would be, as his first arrest had taken three years to accomplish. The story the policeman gave the authorities of the escape was a very pretty one, but the particulars of it were absolutely devoid of truth; but no doubt he is now enjoying his well-earned pension, and his Malay prisoner was saved from a long term of imprisonment, probably owing his life to this incident; for some Malays are like various wild animals, in so far as when deprived of liberty and freedom they pine away and die, and sentences which to a Chinaman mean many years of plenty to eat, contentment, freedom from anxiety and care, to a Malay signify encagement, from which he only emerges to be carried to his grave.

Ambrose B. Rathborne, *Camping and Tramping in Malaya: Fifteen Years' Pioneering in the Native States of the Malay Peninsula*, Swan Sonnenschein & Co., London, 1898, pp. 193–6.

18
The Rickshaw—What to do about it (Hints for Beginners)

ANON

This amusing piece ends on a serious note with a plea for some consideration for the rickshaw puller, whose work was physically exhausting and his living standards poor—with the result that mortality was high.

The rickshaw is said to have been invented in Japan in 1869, by combining large but light metal wheels, the product of Western technology, with a small carriage, to make a very manoeuvrable vehicle superior to the traditional carrying chair. Until displaced in the inter-war period by the motor car (and taxi) the rickshaw, with a puller (usually a Chinese immigrant) between the shafts, was an indispensable form of transport in Malayan towns for the majority who did not own a horse-drawn carriage.

TO many, arriving for the first time out East, the rickshaw has appeared to be picturesque indeed. But when they have been called upon to partake, they have formed an instant desire to be going home again. Properly approached, however, the rickshaw is found to be an animal easily tamed....

A great mistake that many newcomers make is to be in a hurry to get somewhere. As often as not this is due to the fact that the newcomer is a man of business, and he supposes that the business is waiting breathlessly for the man. For four weeks he has paced the deck of his ship, wondering how on earth the firm he is about to join can get on without him and what consoling excuse he is to give for his belated arrival. If you have no better reason for hurrying than this, never hurry. I am not saying that a rickshaw is incapable of a useful turn of speed. I am only telling you that if you expect it to get you to a particular place by a particular time, you expect and perspire in vain. The position is this: you know the destination but you do not know the way. The puller knows the way, but does not know the destination. And there is no community of speech. And, not to put too fine a point upon it, they at the destination don't want you to arrive and you will only be a nuisance, when you do. The right procedure is to clamber into the rickshaw, and leave everything to the puller. He will evolve a much better destination than any you can think of.

There are various ways of getting into the seat, and none of them is correct. You may clamber in over the back, or be dropped in, by a crane, from above. You may back gently in, between the shafts, and fall into position, or you may take a stationary position between the shafts and behind the puller, start the puller off with a sharp nudge and wait to be picked up, as a passive agent, by the seat of the rickshaw when it joins your own. If you adopt the latter method you will find yourself, so to speak, in the well of the court, and you will have to climb up in to the seat of judgment by the best means you can devise. Meanwhile, the rickshaw will pad quietly along, the puller very likely smoking the fag end of a

A rickshaw on the road—one man's effort, another's ease, from J. S. Phillips, *Malay Adventure*, Thomas Nelson & Sons Ltd., London, 1937.

cigarette meanwhile. Personally I always avoid Smoking Rickshaws. Suppose, I say to myself, that the puller's cigarette went out and the puller, in a moment of preoccupation, put his hands into his pockets to feel for a box of matches ...? Where should I be?

An expert passenger will direct the rickshaw by a discreet cough at one point, a rather firm grunt at another point and an arresting sigh at the last. I can never understand the man who can go to more violent lengths. He can have no imagination; for no man could begin even vaguely to conceive himself in the puller's position and not be moved by the most devastating sympathy. I leave it to His Majesty's Judge to decide whether the profession is dangerous, damnable or degrading.... What I refer to is the Cussedness of Circumstances, which is the main element in the rickshaw puller's life: reckless motor cars, indolent and stolid bullock carts, stupid pedestrians, oppressive policemen, everybody wanting rickshaws when it rains and no one wanting them when it's fine, all the troubles of the driver and a horse of a hansom cab or of the chauffeur and the engine of a taxi, and worst perhaps of all, the unspeakable foolishness of the other rickshaw pullers.

To be perfectly candid with you, the sole purpose of this article is to get off a little sermon to you. In all things practise moderation, so far as the rickshaw is concerned. Don't be tiresomely insistent upon a particular destination; really, there is no hurry. Don't be harsh with your puller, for it can do you no good and he has enough to contend with, as it is. Be particularly moderate in overpaying him, for the puller is a man of business like the rest of us and if a good thing is thrust under his nose, he is bound to make the most of it. Overpay, but do not superpay, him. And if you follow this advice your reward shall be great: a really genial, disinterested and engaging smile. When I recall what I have suffered from growlers, hansom cabs, taxis, omnibuses, tube guards and ticket collectors in the West, I feel it has been worth while to come East, if only to enjoy the good rickshaw coolie's smile at the finish.

Anon, 'The Rickshaw—What to do about it (Hints for Beginners)', *Straits Produce*, Singapore, Vol. 2, No. 3, 1 April 1924, pp. 60–2.

19
Hints for Motorists

JOHN ROBSON

The first car, known officially as a 'motor velocipede' and less formally as 'the coffee machine', appeared on the streets of Singapore in 1896, but these outlandish contraptions did not reach Kuala Lumpur until five years later. Cars were not in common use until after the 1914–18 war. In the first years of this century only a handful of enthusiasts experimented in this new method of locomotion.

However, the first decade of this century also saw the beginnings of tourism, marked by an Illustrated Guide to the Federated Malay States, *which was first published in 1911. It included a chapter of 'Hints for Motorists', contributed by John Henry Matthews Robson, who had begun his varied career as a coffee planter in Ceylon, before coming to Selangor to enter its civil service in 1889. In 1896 he had founded the* Malay Mail, *which he edited for some years. Robson, who was prominent and influential as a businessman and newspaper proprietor until the Second World War, was one of the early motoring enthusiasts, and so he wrote the chapter from which the following very practical advice on motoring in Malaya in 1911 is taken. (See also Passage 69 on Robson's relations with Loke Yew.)*

T*YPE of Car*. No special type of car is required for Malayan roads, but the more efficient the water cooling system the better. Two ladies, attended by a native, have travelled through the Peninsula on a 10 h.p. single cylinder Adams car (*vide* 'Autocar' of November 16th, 1907, in which a useful route map is shown). The journey has also been done on de Dions of all sizes, Alldays, Daimlers, Fiats and other cars. De Dion cars are to be met with all over the Peninsula. For two people not overburdened with baggage a 9 h.p. single cylinder de Dion would be a suitable little car, because it is economical to run and well understood in the few local garages. On the whole, however, a more suitable type of car for comfortable touring would be the 14–20 h.p. Siddeley, fitted with dual ignition—a class which would include such cars as the 15 h.p. Zedel, the 15 h.p. Napier, the 14 h.p. Vulcan, the 16 h.p. Humber etc.

Cars with only a 6-inch clearance from the road are not suitable for use in Malaya. There is no speed limit, and the road surfaces are good, but the roads themselves are somewhat narrow, and in many places form an unending succession of sharp corners, which may hide slow-moving bullock carts. In Selangor, Pahang, Negri Sembilan and Malacca there are long and difficult hills to negotiate. An average of 18 miles an hour would be quite good enough for strangers to attempt.

Accumulators can be charged at Penang, Kuala Lumpur and Singapore, but cannot be recommended for use on a touring car in the tropics. Dry cells can be purchased wherever petrol is obtainable. Cars fitted with Bosch high tension magneto seem to give little or no trouble, but dual ignition (magneto and dry cells) is strongly recommended. Cape cart hoods are in general use, but unless there is a lady passenger long journeys are often undertaken with the hood down. A Stepney wheel or detachable rims are necessary, and two complete spare tyres should be carried in addition. A Gabriel horn is recommended, and a spare petrol tank on the footboard may be found useful. A light waterproof cover to place over the car at night in places where no building is available should be included in the outfit. Metal non-skids on the back wheels are a protection against cast bullock shoes, which have very sharp edges. Petrol, costing from 1s. 9d. to 2s. 4d. a gallon, can be obtained in Penang, Taiping (Tate & Co.), Ipoh (Riley, Hargreaves & Co.), Kuala Lumpur (Harper & Co. and Zacharias & Co.), Klang, Seremban, Malacca (a Chinese shop) and Singapore. Strangers will experience no difficulty in finding any of these petrol depots. Repairs can be done at Penang, Taiping, Ipoh, Kuala Lumpur, Seremban and Singapore.

A very few words of Malay will carry travellers all through the Peninsula, but it is advisable to engage a Malay driver or cleaner to assist with tyre renewals etc. He should not be allowed to drive or adjust a strange car. The cleaning will probably be of a somewhat perfunctory nature, but Malays are good-tempered and obliging. Such a man could probably be engaged through Messrs. Young & Co., Barrack Road, Penang, or through Messrs. Wearne & Co., motor-car dealers of Singapore (say $10=£1 3s 4d a week, and expenses). An English chauffeur would have to be treated as one of the party, except in the large towns, and his expenses would be the same. Cars with a long wheel-base are not suitable for Malayan roads; 8 ft. 6 ins to 9 ft. would be found convenient.

Conveyance of Car. Special arrangements would have to be made in London for the conveyance of a car on ocean going steamers by which the passengers themselves travelled. It is better to do this and carry letters to the local shipping agents rather than rely entirely on these gentlemen, who, although personally willing to oblige, might be in possession of printed instructions which would prevent them accepting an uncrated car.

Personal Outfit. Many of the helmets and topis [pith hats] sold in London are quite unsuitable for use in the tropics and serve only to proclaim the stranger! A pith topi with quilted khaki cover and a strong strap to go right under the chin is the best form of head-covering when motoring in Malaya. Such a topi can be obtained at the Army and Navy Stores under the name of 'Cawnpore Tent Club'. A pair of dark glasses will protect the eyes from the glare of the sun and add considerably to the comfort of the traveller.

The most workmanlike outfit for the motorist is a khaki coat and breeches with spiral putties and light boots. The coat should be cut as a Norfolk jacket or tunic with loose military stock collar. A gauze singlet, short thin flannel drawers, thin socks, a Burberry rain coat, and a cap for the evenings would complete the outfit....

A less workmanlike but equally suitable outfit would be a very *light* English coat and waistcoat over a soft fronted shirt with light flannel trousers. The ordinary flannel trousers in use at home would be found too hot and too heavy for Malaya.

Except that a topi is advisable, ladies can gauge their own requirements by remembering the hottest day they have known in England. Local washermen are of the rough and ready order, with emphasis on the rough. Light grey, fawn or mauve colours are recommended in preference to plain white for travelling.

A revolver is not necessary, but there is no harm in carrying one. A licence which is obtainable at any police office, costs only a shilling or two.

J. H. M. Robson, 'Hints for Motorists', from C. W. Harrison, *Illustrated Guide to the Federated Malay States*, Malay States Development Agency, London, 1911, pp. 209–12.

20
Learning to Drive

CHOO KIA PENG

Choo Kia Peng became, between the wars, the leading Chinese businessman in Selangor, and a member of the Federated Malay States Federal Council. He here recounts how, while in the service of the celebrated Chinese millionaire, Loke Yew ('the Towkay') (on whom see Passage 69), he learnt to drive the early motor cars which Loke Yew, like Robson (Passage 19 above), a pioneer of motoring, owned in the early years of this century.

The names of the roads and buildings have been changed in modern times. The car caught fire in Leboh Pasar Besar and the episode of pushing the car uphill back to Kuala Lumpur occurred at the far end of Jalan Damansara. 'Carcosa' was at this time the house of the Resident-General, at which the High Commissioner entertained, when on visits from Singapore.

MR Zacharias brought the small car to Loke Yew's office between the General Post Office and Straits Trading Co. with an Indian driver, and my Towkay said, K. P., let the driver take you for a short run and tell me how you like it. There was no starter. You started the car by kerosine oil at the back and waited for a few minutes until the engine warmed up. The steering again was different, like a Marshall's baton, 2 ft. up and 2 ft. flexible and 1" round.

We started from the office through Market Street, Old Market Square, Old Ampang Street—and the back of the car, where we started it with kerosine, was on fire. I jumped down, rushed to a Chinese shop and asked for an empty gunny rice bag, and with it we extinguished the fire. We cautiously, with some more kerosine, set the match gingerly and after 5 minutes the engine responded to the driver's call. I said to the Indian driver, We are going to take the shortest way home. He drove it through Java Street and took to the Embankment Road. Just at the Government Building, present car garage, he lost his head again and turned left down the Klang River. Thanks to the P.W.D. road metal piled up in cubes, our small car went over it riding astride, with two wheels on the river side and hind wheels on the road side, sitting on top of the metal pyramids. I said goodbye to the driver and walked back to the office. But my boss purchased it, as the incident was not the car's fault.

The second car, from Glasgow, must have been not less than 15 h.p. It was a six seater, two front seats, four at the back, like the old carriages of England, facing each other, and you get up through the back-centre step. And, believe me, the car radiator was supplied with water for cooling from a foot-square copper tank next to the chauffeur's seat, and—in consequence—from our office in Market Street to my boss's house in Batu Road, we often had to stop to give the small copper box a drink when its water boiled....

The De Dion Bouton now I remember was purchased in 1901, after Mr Zacharias' car. I thought then it was a wonder car to be able to go about with benzine and a handle to start it without using kerosine oil burners.... Later on, in the year 1902, it was the same car, driven by Ahmad. The Towkay took me to Tanjong Malim and asked me to watch Ahmad, how the car was driven. We stopped 2 or 3 days at the estate and it was then I was asked by my Towkay to accompany Mr Bernard, the Deputy Conservator of Forests, to examine the 2,000 acre jungle at Ulu Sungai Tinggi. On our way back to Kuala Kubu the Towkay said, K. P., you take the drive now and with Ahmad near you might be able to make it. I did make it without mishap, a distance of about 16 miles. I was elated and in the evening I quietly took the car out at dusk and drove it through Kuala Kubu town and home gleefully alone, and on returning home, at the Kuala Kubu bridge a European, a Mr Bullen, the Town Executive Engineer, said, Kia Peng, you have no light and it is 7.30 p.m. now.

A week after my boss called me to go down to Kuala Lumpur. He said, Will you drive Towkay Chan Sow Lin,

M. L. C., Towkay Loke Chow Kit and me tomorrow evening to Carcosa (King's House was not yet built) to attend the King's Birthday Ball given by Sir John Anderson. Each of the three weighed over 150 lbs. I pleaded my short eyesight was not good for night driving, and he said, We would help you. There was not much help when two of these fat gentlemen sat facing me. I said, I have no licence yet. He said, Get the money from the cashier. So with $10, I went to the Kuala Lumpur Sanitary Board, got a receipt and in a few days received my silver badge, the size of a 50 cents silver, on one side my name and on the other my number—28.

I took them to Carcosa alright, but it was on the return journey I was anxious. It was 1 o'clock when I drove them home. On the left I knew it was Sydney Lake and I kept more on the right and made false turns twice to the entrances of two bungalows on the right coming down. Almost reaching home, not more than 50 yards from the Chartered Bank, I took too much left to get a nice right turn into Market Street. I nearly turned the car over. The left wheel went up a heap of sand, P.W.D. road materials. My second thanks to the P.W.D. as the materials were sand only.

A week after my Towkay said, Your driving is quite good. Now I am visiting my rubber estate (Batu Enjor, 3 miles from Klang, later sold to the Highlands and Lowlands Para Rubber Co.). I shall be going down by train at 9 o'clock and I hope you could get there in time for me. I said, Yes.

Ahmad, in teaching me to drive, also explained to me minutely the mechanical side of the car and its tempers. He said, You are under no circumstances to go long distances without carrying a good spanner with you. The gears often get slack and it will not climb at all, until you have tightened it in a reverse way. In my anxiety to please my boss, I forgot Ahmad's good advice. So the next morning, at 7 o'clock, I left Kuala Lumpur for Klang, without knowing the road, with a large tin of 4 gallons benzine, without a spare wheel or inner tube and the indispensable spanner.

I went to the Railway Station, Damansara Road side, and asked the Indian porter which way will lead me to Klang. He pointed out to me to go straight to Damansara Road. I went straight, without turning left by Travers Road, up the hill and came to the top of King's House Hill and an easy 4 miles downhill all the way, and found the 7th Mile Village. I drove another 100 yards, and that was the end of my Klang Road, a rubber estate facing me and no road.

Then I realised I must have come to the wrong road, so I turned the car back. When I started to climb to the village it simply refused to budge. Then I remembered Ahmad's advice, but it was too late, so—to make the best out of a bad job—I pushed the car up to the village from the left side and kept steering at the same time. I appealed to a Chinese bullock-cart to trail me home. He refused even an offer of $4. Then I enlisted the help of a Malay boy at $1.50, and we both pushed the car. It was hot at noon then. We could only do it for half a mile. Then we managed to get a second Malay assistant. We pushed another mile. At a nearby Malay house we managed to enlist a third helper. We succeeded this time to do two miles and had to enlist a fourth, and thus the car was brought to the top of Carcosa Hill, and to Kuala Lumpur downhill all the way. With the four Malays in the car, I free-wheeled down to High Street, the Federated Engineering Co., and borrowed a spanner from Mr Dearie Russell, a motor-car enthusiast, and the car was right again, and I went to Loke Yew's office to pay off the Malays. It was 5 o'clock. How hot! The Towkay knew of my failure, but he blamed me not, but the car. Perhaps he wanted me to persevere to master the driving.

Choo Kia Peng, 'My Life's Journey', unpublished autobiography, *c.*1953.

Kuala Lumpur down the Years

21
Kuala Lumpur in 1872

FRANK SWETTENHAM

*Kuala Lumpur had its origin as a very small trading post at the confluence of the Klang and Gombak Rivers, a year or two before 1860. In 1857 Raja Abdullah, chief of Klang and the Klang valley, had sent a party of eighty-seven Chinese miners up the Klang River to open tin mines. As the river ceased to be navigable for boats at the confluence (*kuala*), cargoes were loaded and unloaded at that point and carried as porters' loads to or from the mines a few miles on. Yap Ah Loy, who became the headman (*Capitan China*) of Kuala Lumpur in 1868, was the leader of its Chinese community until his death in 1885, by which time it was a much larger town and it had become (since 1880) the administrative headquarters of the colonial State government (established at Klang in 1875).*

In the course of the destructive Selangor civil war (1867–73) attackers three times burnt the village of Kuala Lumpur to the ground, but after each disaster Yap Ah Loy induced his men to return and rebuild it. The Singapore lawyer, J. G. Davidson (see Passage 14 above) had connections with Selangor which brought him to Kuala Lumpur, on a brief visit in 1872, just before it was destroyed for the first time. Frank Swettenham, then a young administrator, accompanied Davidson. His account, published seventy years later, is the earliest extant description of Kuala Lumpur, and the only one we have of the town before its destruction in the war. However, there is reason to think that, as rebuilt after the war in the mid-1870s, it was very like the pre-war settlement.

This is the first of a sequence of passages offering glimpses of Kuala Lumpur at various times down to the middle of the present century. A later group of passages ('Ceremonies and Recreations') provides some additional material on sports and amusements in Kuala Lumpur, particularly in the 1890s.

THE accused was defended by Mr James Guthrie Davidson, principal partner in the firm of Rodyk and Davidson, and then the leader of the Singapore Bar. He was rather a friend of mine, mainly because of our mutual interest in Malays and things Malayan. Therefore, a few days later, he invited me to dinner and told me some particulars about the charge against the Chinese headman. Davidson was convinced that his client was innocent, and that the girl had been smuggled away into the Malay State of Selangor where she was being detained by Chinese of another Society. His intention was to go to Selangor and try to find her and he invited me to join him in the quest. I was delighted to accept the offer provided I could get leave, and rather to my surprise, this was granted. Davidson's firm acted as legal advisers to the Maharaja of Johore—as he then was—and Davidson also knew and had had dealings with Tunku dia Udin—commonly called Tunku Kudin, brother of the Raja of Kedah—who had recently married the daughter of the Sultan of Selangor, and was then living in a fort at Klang on the left bank of the Klang River. Tunku Kudin had persuaded the old Sultan Abdulsamed to appoint him Viceroy of Selangor, because the State was, and had been for years, the war playground of a number of Malay Rajas, whose pastime was fighting and intriguing to gain control of rich districts in Selangor where Chinese, and a few others were mining tin. The main centre of mining was Kuala Lumpur, a purely Chinese village, consisting of two rows of adobe-built dwellings thatched with palm leaves, under the unquestioned control of a redoubtable Capitan China who, having been drawn into the fighting for the protection of his mines and his people had allied himself with the Viceroy. The fort of Klang was seven miles from the mouth of the river, navigable to that point for small steamers, and Kuala Lumpur was twenty odd miles further in the interior, reached either by boat poled up the ever-narrowing river, or through the roadless jungle, miles of it a deep morass.

This was the place where Davidson proposed to take me for a holiday, and where he intended to make enquiries on the chance of finding the missing girl.

We left Singapore in a very small steamer and duly reached Klang, where the Viceroy accommodated and entertained us until arrangements had been made for a boat to take us up river to Kuala Lumpur. The journey took three days, rowing and poling, and we were welcomed by the doughty Capitan China Yap Ah Loi and his friends who, in the evening, entertained us to a great dinner, my only recollection of which is that many Mexican dollars had been turned into spoons and forks for our use. It was very thoughtful, but the forks, being pure silver, bent under the smallest pressure and had to be constantly straightened in order to carry the food.

With the exception of the Capitan China's own house—which was more pretentious and more solidly built—the place consisted of thatched hovels with earth flooring, some of them unoccupied. The next day, while Davidson was making his enquiries, I wandered round Kuala Lumpur and went into what appeared to be an empty hut: it was quite empty, except for a dead Chinese, with a bullet hole in his chest, who was sitting on the red earth floor with his back against the wall.

In the afternoon we walked to some tin mines a few miles away, and Capitan Ah Loi insisted upon sending some Chinese warriors as a guard. Of course there was only a path through the jungle and we had to walk in single file. The young warrior in front of me—clothed in the shortest of shorts and a large palm-leaf hat—had a loaded ten-chamber revolver hanging by a piece of string from a stick carried over his shoulder. As the weapon bumped continually on his bare back during miles of walking over roots and jumping across streams, I saw that the solitary figure sitting on the red earth of the empty hut could be accounted for.

A day or two later Davidson told me that his quest had resulted in failure, and we decided to walk the greater part of the way back to Klang. It was a twelve hours' effort and very strenuous and unpleasant at that, for there was no path and much of the distance we travelled up to our waists in water. Torn by thorns, poisoned by leech-bites, and stung by scores of blood-sucking insects, the struggle was one long misery. Of course we had a guide, otherwise we must have perished, as had been, a short while before, the fate of the Viceroy's 'foreign legion', engaged in Singapore to deal with his rivals, the fighting Rajas. At last we reached the bank of the Klang River, found a boat, and were rowed the remaining short distance to Klang. As a result of that walk, Davidson was so distressed and sore-footed that he could do nothing for two or three days, by which time it was necessary—for me at any rate—to get back to Singapore. There was no steamer, nor prospect of one, so, with considerable difficulty, Davidson got a small native sailing boat, with three or four men, to make the journey. The distance is about two hundred miles, and when we cleared the mouth of the Klang River and its islands, we found that our so-called crew were none of them really sailors and that we should have to do most of the directing and steering ourselves, but we hoped that by hugging the coast we should manage to reach our goal. We did manage, but the journey took longer than we expected; there was nothing to eat but rice, salt fish and some durians. That was the only occasion on which I persuaded myself to eat a durian.

Frank Swettenham, *Footprints in Malaya*, Hutchinson, London, 1942, pp. 20–1.

22
A Yank in Kuala Lumpur in 1878

WILLIAM HORNADAY

William Hornaday, a young American zoologist (later Director of the New York Zoo), made a two-year expedition to India, Ceylon, Malaya, and finally Sarawak, to collect specimens of wild animals for exhibition (in preserved form) in American and other museums. He made a brief visit to the interior of Selangor in company with H. C. Syers, head of the Selangor police, who shared his enthusiasm for tracking and shooting big game.

Hornaday's book, recounting his experiences in vivid style, went through many editions.

AFTER passing two or three clearings, we reached the top of a long, steep hill, and, at its foot, Kwala Lumpor lay before us, on the opposite bank of the river Klang, here reduced in size to a narrow but deep creek. A sampan came across to ferry us over, while our ponies swam beside it, and at 5 P.M. we were at our resting place for the night.

All along the river bank, the houses of the Malays stand in a solid row on piles ten feet high, directly over the swift and muddy current. The houses elsewhere throughout the town are walled with mud, and very steeply roofed with attaps (shingles made of nipa-palm leaves), so that a view of the town from any side discloses very little except high, brown roofs slanting steeply up. In the centre of the town is a large market where fruits, vegetables, meats and various abominations of Chinese cookery are sold. The vegetables are sweet potatoes, yams of various kinds, beans, melons, cucumbers, radishes, Chinese cabbage, onions, egg-plant and 'lady's fingers'. The fruits were the durian, mangosteen, pineapple, banana, and plantain, oranges (of foreign growth), limes, 'papayah', and other small kinds not known by English names.

In the centre of the market-place are a lot of gambling-tables, which, a little later in the evening, were crowded with Chinamen earnestly engaged in the noble pastime of 'fighting the tiger'. The principal streets are lined with Chinese shops, and are uniformly clean and tidily kept. The streets inhabited by the Malays can be recognized at sight by the accumulation of dirt and malodorous rubbish, and the dilapidated appearance of the houses.

We went straight to the house of the Captain China [Chinese headman] (pronounced Cheena), the man of importance in the district, who is governor of the Chinese in every sense of the word. His title is Sri Indra Purkasah Wi Jayah Buktie ('Fair-fighting Chief and Hero'), and his name, Yap Ah Loy, commonly called by Europeans the Captain China. In return for his services to the district in opening new roads and preserving good order, with his own police force, the government allows him a royalty of $1 on every bhara (which equals three piculs, or four hundred pounds) of tin exported, and from this source, and also from his eleven tin mines, he is said to be the wealthiest man in the territory. He has in his employ sixteen hundred and twenty-seven men, and entertains at his house, in true European style, every white man who visits Kuala Lumpur. Unfortunately he was absent at that time, but his people received us quite as if he had been there, and made us comfortable with a fine dinner, an abundance of excellent champagne and good beds.

The next morning, while in the largest Chinese store in the place, buying provisions for our stay in the jungle, we struck a bonanza. We found Mumm's champagne for sale at sixty cents a quart, and India pale ale at fifteen cents per pint! How they ever managed to sell either at such ridiculously low prices we could not understand, and, to ease our consciences before victimising the dealer, we told him he must have made a mistake in marking his goods. No, that was the price, and we could have all we wanted. It would have been flying in the face of a kind Providence to have neglected such an opportunity as comes but once in a lifetime.

Engaging the strongest coolie we could find we loaded him with champagne (at sixty cents per quart!), and marched him ahead of us into the jungle. It was the proudest moment of my life. I may never strike oil, or gold-bearing quartz, or draw a prize in the Louisiana lottery; but I have struck Jules Mumm's best at sixty cents a quart. My only regret is that I did not fill a tub and take a bath in it, for champagne is the only artificial drink I really like.

William. T. Hornaday, *Two Years in the Jungle: The Experiences of a Hunter and Naturalist in India, Ceylon, the Malay Peninsula and Borneo*, Charles Scribner's Sons, New York, 1885, pp. 315–16.

23
A Royal Visit to Kuala Lumpur in 1879

BLOOMFIELD DOUGLAS

The ruling dynasty of Selangor were Bugis (originally from the Celebes) who had established themselves in the mid-eighteenth century in control of the coast of Selangor. The scattered Malay villages of the interior were settled mainly by Sumatran immigrants, who came to mine tin. A quarrel between the Sultan's Bugis headman at Kuala Lumpur and local Sumatrans had precipitated the civil war (1867–73) though it was by no means the only cause of the conflict.

No Sultan had visited Kuala Lumpur since its foundation twenty years before (Passage 21 above). At the time of this visit, Sultan Abdul Samad was in his seventies. Both he and the British officials who came with him were apprehensive as to the reception which his subjects in the remote interior might accord him. But this passage from the diary of Bloomfield Douglas (Resident of Selangor, 1876–82) records the enthusiastic welcome given to the Ruler. The British administrative headquarters in Selangor was moved to Kuala Lumpur in the following year, but at the time of the royal visit in 1879 Yap Ah Loy and the Malay headmen were still in control of Kuala Lumpur and the surrounding mining camps.

The imperfect punctuation of the author's text has been retained but where words seem to have been omitted, conjectural replacements have been supplied.

Wednesday 7th

UP early getting ready for a start. The Sultan quite bright and active, looking forward with pleasure to his trip to Damansarah and Kwala Lumpor.

At 9.15 we embarked and mustered some 65 persons aboard the Abdul Samat, an escort and horses had been sent on yesterday. Arrived at Damansarah at 11.30 against a strong ebb. The Sultan's party consisted of himself, Tunku Musah, Rajahs Sah and Slayman, Tunku Panglima Rajah, Raja Mat of Tanjong Gamok, Tunku Abdul Rahman, Inche Ambok, 2 female attendants and a host of followers.

I took two of my ponies and Messrs Syers and Turney took theirs. Mr Daly's horse refused to go in the boat.

My party consisted of myself, Mr Daly, Mr Turney, Mr Syers and six of my guard.

We found five more ponies at Damansarah, a large sedan and some mountain chairs with bearers so with bullock drays we had good means of transport.

The heat was something fearful, the coolies complaining most bitterly. We stopped at the Datu Mangkoos and had some refreshment. At the 6th milestone at the Klang Company's plantation we made another halt and there I introduced to the Sultan Seyd Sallim a brother-in-law of Seyd Mahomed Alsagoff and Mr Limke's successor. He was invited by the ruler to accompany him to Kwala Lumpor and he did so.

There was some little hitch when the bullock drays arrived at the end of the made road about 3 miles from our destination—I was obliged to press into service 30 or 40 of the Captain China's road coolies. On the rising ground above Kwala Lumpor large parties of Malays, Selangor, Mandaling, Menangkaboo men etc. met the cortege, paid homage to the Sultan and then fell into the rear. As we neared the town we could not fail to admire the beauty of its site, the nice appearance of the bridge and decorations so profusely exhibited in honour of the Sultan's visit.

At the foot of the first rise we met a number of Chinese towkays who saluted H.H. and took their place in the procession. On arriving near the bridge the Sultan left his sedan chair and took my hand, we had previously sent the horses off the line of the road and proceeded on foot—I felt the poor old man's hand tremble with excitement but his face was lit with pleasure and I think some degree of pride at the reception accorded to him.

The Captain China and Towkay Ah Shak of Petaling next appeared in mandarin, or at any rate very gorgeous array and gave a very welcome greeting to H.H. We then advanced all sorts of barbaric music clanging around the Sultan, the crowd which was very dense shouting and cheering at the top of their voices and then we crossed the bridge and entered the shed leading from the river to the Captain's extensive premises. Here was stationed a strong guard of honour under a canopy or roof of white cloth, adorned at the sides with many gorgeous banners, a salute of I don't know how many guns was fired, and we passed beyond the ordinary entrance to the Captain's house and the market place to the shed leading towards the [front] half of it. We stopped at a very nice gateway and entered the Capitan's campong all nicely decorated and certainly on this occasion most scrupulously clean. There was a strong bamboo fence all round the quarters prepared for the Sultan to which we were glad enough to see him safe into. Sentries were placed at all the gateways and I took my leave of the good old Sultan congratulating him on his safe arrival and the strength with which he had borne up against the fatigues of the day better indeed than most of the people, who one and all complained greatly of the heat. The day was a scorcher and no mistake....

Thursday 8th

Up early and heard that the Sultan like ourselves had not spent a very good night—visited the fort and saw Rajah Shaban and Datu Satee, a great man among the Menangkabu men it is stated he has a following of nearly 1,000 men, whom he asked permission to take before the Sultan....

About 1 p.m. there was a large concourse of people who tendered their homage to the Sultan, some thousands were present—I should say from 5,000 to 7,000 persons were congregated in the town; some estimated the number at 10,000 but this is excessive in my opinion. The Malays first paid their respects then the Chinese. The Sultan wore his state uniform and a very strong guard of honour under arms attended under the command of Mr Syers, every possible precaution was taken against treachery or the [folly] of any of the half-witted opium smokers—but who could injure such a good creature as Abdul Samat? And so all passed off very well without a hitch or the slightest appearance of anything but normal devoted loyalty to the Sultan....

After this the Sultan joined us and we witnessed the races and sports. Everything went off very well indeed. There were two greasy poles, one assailed by Malays and the other by Klings.

The crowd was very dense, but every one was in good humour and appeared to enjoy the fun. Raja Musah I am sure never laughed so much in his whole life, at any rate I never saw him doing anything but grin and grunt approval before, on this occasion he was convulsed with mirth.

The Malay women of course did not appear but when the fun at the greasy poles was at the highest I saw lots of bright eyes peeping from under the uplifted Singkups [blinds] of the atap roofs supported in some cases by well shaped fair hands.

The Klings were the winners, night came and so the fun ended. We were all pretty tired.

I saw the Sultan before dinner in the evening he went to the Chinese wayang, attended by the Capitan China, Mr Syers and Mr Turney.

Friday 9th

At an early hour we accompanied the Sultan and his attendant Rajahs and suite to the fort where he placed the first pole of the new building, Court house and rest house combined with Malayan ceremony a wreath was fastened to the top of the pole and it was then smeared with 'Meneak Atal',* a short prayer was said, the pole raised and adjusted in its position and the Sultan addressed a few words to those present, including 6 light sentence convicts.

He said in thus raising the pole of a new hall of justice he hoped its use in that respect would be limited, that it was better to work hard and even live poorly than to commit crime, small faults such as all men may be guilty of could be passed by but men breaking the law and committing serious crimes must expect punishment, a second offence would always be more severely dealt with.

Having previously advised H.H. to do so—he called the 6 convicts before [him] and said, as their cases had not been grave ones, that most of their short sentences had nearly expired he would pardon them. This took them very much by surprise, but they all dropped at the Sultan's feet and I believe expressed in their act of homage the gratitude they felt.

W. Bloomfield Douglas, Diary entries 7–9 July 1879. (Most of this passage has been published, with slight changes, in the *Journal of the Malaysian Branch of the Royal Asiatic Society*, Vol. 48, Pt. 2, 1975, pp. 32–5.)

**Minyak hartal* was a paste of yellow arsenic used for its supposed magical properties.

24
An Aussie in Kuala Lumpur in 1883

AMBROSE RATHBORNE

This is a second extract from Rathborne's reminiscences (see Passage 17 above). Although he dates this visit to 1883, the building regulations to which he refers in the last paragraph were made in 1884.

As a result of a mining boom which began late in 1879, Kuala Lumpur had grown rapidly; in 1884 it had a population of about 4,000. Fires could easily spread through the wood and atap (palm thatch) buildings of Yap Ah Loy's Kuala Lumpur, as they stood along either side of narrow lanes. Kuala Lumpur was almost completely destroyed by fire in 1881. The insanitary condition of the town in 1882 was described by the Governor, during a visit, as 'pestilential'. Hence the official policy, accomplished during the mid-1880s, was to get the town, of some 500 houses and shops, entirely rebuilt a street at a time, with brick walls and corrugated iron roofs for the buildings spaced back along wider roads. By 1887 there was only one building standing which was more than five years old.

ADJOINING Sungie [Sungei] Ujong to the north lies the State of Selangor, and in 1883 the only means of communication between the two places was by a jungle track, which some distance from Seramban lost itself in a muddy stream until the foot of the S'tul [Setul] range of hills was reached. After surmounting these and descending on the other side, the valley of S'tul was passed, and amongst the few houses dotted round the paddy fields were some inhabited by men from Karinchi in Sumatra, who have the reputation of being able to transform themselves at will into tigers, a superstition firmly believed in by their neighbours, who hold them in some dread and awe.

The next inhabited place reached was the isolated village of Brennang [Beranang], consisting of two or three wayside houses of the usual flimsy kind, easily built, and as lightly abandoned should the dwellers care to leave them and move elsewhere. Each house had a rough shelter for passers-by, in which hung sundry bunches of shrivelled plantains for sale, and which were generally occupied by half-naked children, who made these sheds their playground. Here the traveller rested awhile before attempting to cross the swamp, full of rank grasses and rushes, and waist-deep in mud, that stretched across his path.

Just as night was falling the river of Samunieh was reached, and in the village beyond the wayfarer felt thankful that half the journey had been achieved, and that the next day would bring him to Kwala Lumpor. Not, however, that the second day's journey was any better than the first; more tedious, if anything, was the narrow jungle path, full of holes and roots; the paddy fields of Cadjan [Kajang] and the river at Cheras had all to be crossed, as well as innumerable streams and spurs and swampy gullies, before the mining camp of Pudu was reached, an outskirt of Kwala Lumpor, and a settlement of Chinese miners, who were all busily employed in 'winning' the tin with which the valley abounded.

Kwala Lumpor is the chief town of Selangor, and the principal Government offices are located here. It is situated at the mouth of the Gomba river, a tributary of the Klang, the latter being the chief river of this part of the country. The houses of the Government officers were pleasantly situated on the adjoining hills overlooking the town, roads were in course of being laid out, order was being kept by a small body of Malay police, and Mr (now Sir) Frank A. Swettenham had lately been appointed the British Resident of the state; and its rapid development was in large measure due to the policy adopted of improving the means of communication to enable the Chinese miners to transport their supplies at a reasonable cost, and also to the encouragements to settle that were extended to the Malay immigrants from foreign states, who readily availed themselves of the advantages given.

The immediate neighbourhood consists of a fine fertile basin of flat country forming an amphitheatre, surrounded on the east by the mountains of the main range, and on the other sides by subsidiary spurs. The hills here are bolder and the valleys more extensive than further south, and these characteristics become even more marked in the state of Perak, further to the northward.

The town itself already presented all the appearances of a

prosperous mining centre; the streets were littered with bricks and timber, for substantial structures were fast taking the place of the flimsy wooden houses so liable to catch fire and be destroyed. It is no unusual occurrence for a whole village, which had quickly sprung into existence owing to some great influx of Chinese miners to the neighbourhood, to be entirely devastated and laid waste by fire, a layer of ashes and a few badly-charred posts here and there being all that was left of what had been but a few hours before a flourishing little centre of trade. In the towns, of course, the destruction was on a bigger scale, and the opportunity was taken advantage of by all the bad characters to lay hands on and steal what they could; rioting and fighting also created a new danger, and made the confusion worse. The better houses were formerly built with mud walls, and over the ceiling there was a layer of earth. On the first alarm of fire, a hurried rush would be made by the inhabitants to close the doors of their shops in order to prevent their contents from being looted. The owners of the wooden houses nearest the conflagration were busily employed in carrying what they could of their goods to some place of safety. Those in the mud houses simply sat inside and patiently awaited events, in comparative safety if the conflagration was not too fierce; for although the light roofs overhead were burnt and destroyed, the contents of the shop were but slightly damaged unless the fire gained an entrance through the wooden doors or windows, in which case the building would be entirely gutted. On no consideration would those inhabitants who were somewhat more remote from the fire help to extinguish the flames or open their doors, and the only way to gain an entrance was by bursting them in. As an instance of this, I have seen the roof of a shop catch fire from some spark that had blown on to the thatch unknown to the inmates within, who obstinately refused admittance to those outside endeavouring to enter, so that they might get on to the roof and put out the flames. Shouting and hammering were of no avail, and there was nothing to be done but to break in the door with an axe, when the Chinese occupants were disclosed crouching down and awaiting events in dumb stupidity, seemingly paralyzed by the dread of being robbed should they open their doors and by the fear that the fire after all might reach them. Then in turn house after house had to be broken into, and the inmates compelled to fetch water to throw over their roofs in order to prevent stray sparks from igniting the palm leaves with which they were thatched.

It was to prevent the destruction of property and its concomitant evils, that directly a village became prosperous and of sufficient importance, orders were given that within a defined area all the shops were to be built of brick before a certain date. This decree had been issued at Kwala Lumpor just before my visit, and accounted for the roads in the parts affected by the order being littered with building material.

Ambrose B. Rathborne, *Camping and Tramping in Malaya: Fifteen Years' Pioneering in the Native States of the Malay Peninsula*, Swan Sonnenschein & Co., London, 1898, pp. 105–9.

25
Shopping and the Social Round in Kuala Lumpur in 1899

ETHEL DOUGLAS HUME

Miss Hume's visit to Malaya lasted only a few months (see Passage 10) but, as the sister of a government official, she was effortlessly absorbed into the European social world. She here describes going shopping in Kuala Lumpur with a friend and then the evening's amusements.

ONE after another we passed the pawnshops, where we caught glimpses of fascinating jewellery and curiosities of all sorts. 'Hammer, hammer, hammer', resounded from the tin shops where the metal was being welded into all sorts and conditions of designs. Here, in front of a silk shop, a little shrivelled-up owner sat smoking his opium pipe; there, outside a barber's, were Chinese having their heads shaved and their ears cleaned with silver instruments, this all-engrossing occupation being part of a barber's task. Some of the shopmen were having tiffin rather early, and were sitting round tables, and with chop-sticks and china spoons partaking of greasy fluids with submerged lumps of pork, finishing with smelly delicacies, such as decomposed prawns, sharks' fins, fishes' maws, and swallows' nests.

We were rather relieved to step into a darkened interior out of the glare and heat. We were at perfect liberty to examine all the shopman's possessions, but when it came to the point of making a purchase the difficulty began, as an anxiety to sell is considered 'bad joss' by a Chinese salesman, and the fact that he is supposed to wish to part with any of his belongings appears to meet with his profoundest contempt. 'T'ada' (not got), Mrs Freshcombe and I were assured repeatedly, while the object we asked for stared us full in the face....

It was not only so difficult to choose things, but harder still when it came to the question of price. To begin with, in many of the shops there were workmen nailing, hammering, or otherwise making a noise. They had to be relieved of their instruments before it was possible to hear anybody speak. When the shopmen at last made themselves audible it was only to mention some sum four or five times the value of our would-be purchases; and though they had not the least expectation of getting what they demanded, the toil of diminishing prices was not reduced by that fact. It was particularly distracting when statements were corroborated on the abacus, as ordinary arithmetic has always been my bugbear, and it was positively overwhelming to be confronted by Chinese sums. Besides, I never could see the connection between a counting-board and an exorbitant price. Mrs Freshcombe grew rather nervous when I simplified transactions by halving or quartering the amounts, according to their requirements, laying down what I thought proper, and walking away with my purchase, regardless of the cries of 'Mem! Mem! Hey! Hey!' with which I was pursued....

Mrs Freshcombe was also the first to introduce me to the Europeanised Chinese shops, where the salesmen wore western garments and straw-hats or bowlers above their queues. European goods were chiefly supplied at what were supposed to be fixed prices, but it was quite as tedious to deal there, as the assistants scorned Malay, and in slow measured tones gave vent to the most laboriously grammatical and roundly finished English sentences, which took so long to utter that I had always forgotten the beginning by the time the speakers arrived at the end.

Such a strain on the memory made the afternoon siesta more than ever appreciable after a morning of noise and dust in the native town. It was a mercy to find Russian-bath conditions allowable for a short time in the Russian-bath atmosphere, and to feel oneself free to wear a Russian-bath costume, and stare lazily at the drowsy scenery which seemed to yawn and blink in the afternoon heat. But the momentous sixty minutes from five to six had a tiresome habit of arriving precipitately.... From five to six you drive, leave cards, play golf, tennis, or croquet, and then it is suddenly dark, otherwise you would certainly need to change again. If you are neither vain nor nervous of chills you may appear in a crumpled state at the Club, and play cards or study the papers under a punkah. Dinner is a late function generally, though, of course, Chinese servants can serve it to you at any time. They could dish it up for you in ten minutes if you were suddenly filled with a desire for it at eleven o'clock in the morning. It appears a matter of supreme indifference to them if a dozen guests arrive unexpectedly. If your own household goods or edibles are insufficient, the accommodating customs of the East allow your domestics to borrow from your friends. Your belongings will be loaned in their turn whenever your neighbour's needs so require; and you must never at some one else's dinner-party be startled to see your own crest upon the forks and spoons.

Ethel Douglas Hume, *The Globular Jottings of Griselda*, William Blackwood & Sons, Edinburgh, 1907, pp. 46, 54–8.

26
Kuala Lumpur in 1902

NG SEO BUCK

In the post-war period Kuala Lumpur began to rediscover its history, mainly through the reminiscences of a few elderly people, such as Ng Seo Buck, who had spent half a century or more in the town.

When the article from which this passage is taken was written in 1954 many of the streets of Kuala Lumpur still bore the names of former colonial administrators, which have been replaced in more recent years. The High Street (Jalan Tun H. S. Lee) is probably the oldest street in the town, taking its name from the fact that it ran along a ridge of high ground above the expected river flood level. Other streets and open spaces still exist as Jalan Cheng Lock (Foch Avenue), Jalan Hang Kasturi (Rodger Street), Leboh Pasar Besar (Market Street), Merdeka Square (The Padang), and Jalan Tuanku Abdul Rahman (Batu Road), as well as thoroughfares such as Jalan Ampang and Jalan Petaling which are easily recognized as Malay forms of their original names. The temple is again described in Passage 51.

LEE RUBBER BUILDING was then the site of the Capitan China's garden with his Court of Justice adjoining it. The shop houses opposite were the residence of the Capitan China and his seven or eight wives. There was no through traffic in High Street in those days. What is now Foch Avenue was the railway track on which Singapore mail trains were run. At one and the same time, if a train happened to pass, traffic in Rodger Street, High Street, Petaling Street and Sultan Street was held up by railway gates. Luckily traffic in those days was not heavy—a few rickshas, a few bicycles and perhaps one or two horse gharries on either side of the railway gates.

Another landmark of the town is the Chinese temple in High Street. Hemmed in on all sides by shop houses in High Street, Pudu Street and Rodger Street this temple, known as Soo Ya Miow, is now obscured from view. Before the shop houses were built, it would catch the eye of even the most sophisticated visitor. Once in seven years a grand thanks-offering service was held here. Kuala Lumpur was then on holiday, especially for those who had faith in the many wooden idols within the temple. Kowtowing to these monuments created by a mind destitute of cultural common-sense, the devotees came to ask for health, wealth, longevity and love charms, a typical example of Chinese culture debased by superstition.

Around the corner is K.L.'s boast—the Central Market. Of course it is a very recent structure. Delapidated, filthy, vermin-stricken cow sheds stood on the site before this modern, imposing structure was built. One night, a friend who was a perfect stranger to this town, while driving past it, asked me if it was a medical college! Medical study was at the back of his mind.

Here is the Mercantile Bank and here stood the dwelling houses—three in a row—of one of the descendants of Capitan Yap Ah Loy. The one at the corner of Rodger and Market Streets was a druggist shop in which medicine prescribed by the gods of the High Street Temple was dispensed. They alone knew how to decipher the code used in the prescription printed on yellow paper.

On the site of the Hongkong Bank across the road, stood a small shrine known to all as Datoh. I think it was the grave of a mystical Malay and most Chinese turned it into a place of worship.... The Spotted Dog* was housed in very much smaller buildings thatched with attap. It sheltered the Prince of Wales—now Duke of Windsor, in 1922, when that august personage was on his goodwill tour. The Padang would be turned into a Lake when the Klang River swelled its banks.

There were very few places of amusement in those days. The theatre hall in Petaling Street—where the Madras Theatre now stands—and the Sultan Street Theatre Hall, replaced by the Rex, were vermin-stricken halls in which Cantonese Wayang played year in and year out. Occasionally a Teochiew Opera or a *bangsawan*** would give a fortnight's performance in either of the halls. Men who could afford it spent most of

*See Passage 79, page 257.
**See Passage 76, page 245.

their leisure in the licensed opium dens, or in the licensed gambling booths in Petaling Street, High Street, Ampang Street and Batu Road. All sorts of chance games were indulged in ranging from *fan-tan*, dice-throwing, *p'ai-kow* to *chap-ji-ki*. Those who made money spent it in one of the licensed houses of ill-fame many of which, including Japanese brothels, were situated in Petaling Street which was a veritable Yoshiwara.

What were the means of transport? For short distances there was the *jinricksha* and for distances over five miles there was the horse gharry—a sort of big box on four wheels drawn by a horse not much more active than Don Quixote's Rosinante.

The first bicycle—a penny-farthing—was introduced by Bachi, a son of Capitan Yap Ah Loy. His Hainan boy rode it and when he did so, many small street urchins would run after it. You will smile when I tell you that His Lordship Justice Sercombe Smith had to come to office in a two-man-power single-seater ricksha, with one to pull and one to push. This sight was most impressive because the puller and the pusher were clad in khaki uniform with red fringes and they had *towchang* [queues] coiled round their heads.

In days that are dead but not forgotten, K.L. streets were paved with laterite. During the dry season and on a windy day almost every one wore khaki clothes. The streets were illuminated by flickering kerosene oil lamps and most streets had their stand pipes from which the inhabitants drew their daily supply of water. Prisoners, with chained feet and guarded by an armed policeman, swept the streets early in the morning and semi-nude Indian labourers did all the road repairs. In the shopping quarters very few European ladies could be seen. When they did come out they were clad in Victorian dresses, fully veiled and gloved with their skirts trailing along the ground. Chinese women wore black silk dresses or blue; white was taboo because it was a sign of mourning; Japanese women donned their highly coloured kimonos with a big bow behind and were veritable Madame Butterflies; Malay women were '*ber-tudong*' (veiled). On social grounds the East and the West were miles and miles apart. Most Government officials and business men wore closed coats with high collars, known as '*tutup*', and tropical helmets.

Children went to school bare-footed and in the primary classes boys had to learn English and colloquial Malay side by side.

Ng Seo Buck, 'Some Recollections of Kuala Lumpur Fifty Years Ago', *Malayan Historical Journal*, Vol. 1, No.1, 1954, pp. 29–31.

27
Kuala Lumpur under Water in 1926

RICHARD SIDNEY

Sidney (see Passage 16) here gives an eyewitness account of the Kuala Lumpur floods of December 1926. The town had suffered periodic devastating floods from Yap Ah Loy's time, due to heavy rainfall upstream in country which had to some extent been denuded of cover by tin-mining. The eventual cure was to straighten the course of the Klang River, and to confine it within raised banks.

THE town was much more flooded on this Monday morning; and now lightheartedness had given place to frantic preparations in the attempt to stay the encircling waters. Rodger Street had nearly six feet of water, and I was sorry for the inhabitants of the houses who occupied the bottom floor and did not own the first storey as well. It was a hot, sunny day and the sights seemed all the more incongruous in consequence—and I was glad to have taken the precaution of carefully vaselining my legs, for although the water was warm to wade in, yet the sun might soon have blistered the bare flesh. During my journey I saw some queer sights, and near Kampong Attap all except the roofs of the houses were completely submerged and only the heads of some scarlet canna flowers could be seen sticking forlornly out of the waters. The Police were using their field to swim on; while as I looked back to my own bungalow it seemed a forlorn Noah's Ark—which alas! would not float,—while on the verandah I noticed boys moving about. A Malay assured me that it was all due to a dragon which 'had been sleeping a thousand years, and now had awaked and was wriggling his way to the sea!' It was even reported in the local press that one Malay saw it! But there were many philosophers: 'In a quiet corner' near the Town Hall, 'close to the road leading to the power station, an elderly Tamil was standing knee-deep cleaning his teeth in a rather abstracted fashion and making use of the flood for this purpose.' This from the Flood Edition of the *Malay Mail* which had had to be printed by hand because water had flooded the power station. Incidentally the Offices of the paper were flooded and the damage took some days to rectify.

The Court of Appeal was marooned, and when the adjournment for lunch was made, it was found impossible for either judges or counsel to get away. One heard that all the learned gentlemen finally escaped in rickshaws and got very wet, and the spectacle of a Judge of the High Court fleeing in this undignified manner must have put quite a strain on the administration of justice! Another party hired a bullock cart, 'but when they got into the water they found the bullocks trying to swim, and there was danger of the whole crew being precipitated into the flood!' I was told.

Even the railway was now affected, I found, and under the bridge near the M.S.V.R. headquarters the water rushed through the carriages, while the railway goods sheds had become pavilions standing in a lake.

I was so sorry for the small shop-keepers, those in Java

The KL floods of 1926—views of the present-day Bangunan Sultan Abdul Samad and Jalan Tun Perak, from R. J. H. Sidney, *In British Malaya To-day*, Hutchinson, London, 1927.

Street especially were having a bad time—the water even rising above their wooden counters, and there being no place (they are nearly all cloth dealers) for them to store their goods in safety.

I met a policeman and he told me the difficulties they had experienced in rescuing squatters in certain parts of the town. The bridges had been closed to heavy traffic, and the stranded people were afraid to trust themselves to boats. But this was the only way, and one might have seen boats conveying household goods and people away from houses now quite unsafe....

As there was no school it seemed well to take a camera and see what sights there were in the town. It was a dull day, and by putting on stout shoes, wearing shorts and no socks, one might wander and wade freely and be independent of motor car or rickshaw. Looking back over the hedge from High Street it was very curious to see the school buildings standing in a lake, while my bungalow seemed to be completely cut off....

Everywhere crowds of people, and one fraternised as if it were the beginning of War. Boys, usually so shy, accompanied me and were interested in helping me take pictures. The water was coming into High Street and already in one shop there was a foot of it, a Chinese carpenter sat above on a stool looking at it philosophically and making no attempt to put his goods in a safer place. But this was not universal—many traders were taking frantic precautions against the flood: nailing up wooden steps in front of their doors, some even building cement doorways. But the water attacked on both sides, and if you kept it out in front it came in behind.

All the bridges were still passable, but they were feeling the strain considerably. The ordinary river channel had long ceased to be sufficient to carry off the surplus water madly rushing down at us from the distant hills, and the roads parallel with the river absorbed much of the overflow. Perhaps the fine cricket field in front of the Selangor Club was most worth seeing: here was a vast lake bounded by the gables of the Club house behind which rose a palm-studded hill; while opposite was the stately Federal Secretariat and the Post Office already deep in water. In the distance could be seen St. Mary's Church now well awash, and a peep inside revealed the pews floating sadly about.

Many people pretended to disregard the flood, and quite certainly they were not going to wade through the dirty waters! Motor cars splashed through the shallower parts, but got badly stuck in some places; while it was quite customary to sit on top of the back of the seat in a rickshaw and let the puller do his best. As the waters deepened so it became necessary to use boats, and the Post Office showed great enterprise in bringing up *sampans* from Port Swettenham: *they* must try and continue business—flood or no flood!

R. J. H. Sidney, *In British Malaya To-day*, Hutchinson, London, 1927, pp. 252–3, 258–9.

28
The British Leave Kuala Lumpur in 1942

IAN MORRISON

This account of the situation in Kuala Lumpur at the time of the rapid British withdrawal in January 1942 was written by the correspondent of The Times *reporting on the campaign in Malaya.*

THE Japanese must have marched into Kuala Lumpur on January 12th. Most of the previous morning and afternoon I spent in the city together with Til Durdin of the *New York Times*, Gilbert Mant of Reuter's, and Bill Knox of the *Sydney Daily Mirror*.

Most of the British forces in north Malaya had already passed through the city on their way south. There was still a small holding force some fifteen miles to the north. In the city itself demolition squads were blowing up the few remaining bridges. Indian sappers, with pneumatic drills, were boring into the road. The little boxes of white gelignite were stacked on the side. Occasionally there would be a loud explosion. One large iron bridge lay twisted and torn, with its girders in the river and the broken water main gushing its contents into the river too. On the outskirts of the city there were two or three high columns of black smoke—they had been a feature of Kuala Lumpur for several days past—as some remaining stocks of rubber were destroyed. We visited one such fire. The latex was burning fiercely, giving out such heat that one could not go within fifty yards, sending an enormous mushroom of inky smoke straight up into the air. The manager of this estate, an Australian who had been in Malaya for many years, had everything packed up in his car and was just about to leave for the south. The stocks of rice from the godown were being distributed to the Indian and Chinese labourers. Two Indian clerks were keeping a tally. It was quite orderly. Each labourer would have enough rice to keep him for at least two months. In the processing plant next door to the godown all the machinery had been smashed up. The old Chinese who for over twenty years had driven the engine that provided power for the plant seemed utterly bewildered. He himself had had to take a sledge-hammer and damage beyond repair those precious rods and valves and gadgets which he had oiled, greased, tended, watched like a father for so many years that he knew them better than his own children. The manager himself was heartbroken. He was a man who obviously took a tremendous pride in his work. It was the best estate, he told us, for miles around. Look at those young trees over there, never yet tapped, what beautiful condition they were in. He looked after his labour corps well and had never had any trouble with them. They were contented and worked hard. It was obvious that he hated having to leave all these people whom he had come to love, for whom he felt responsible. He was also having to leave his three Chinese servants. He could not take them and their numerous families down to Singapore with him. They would continue to live in the house and look after their master's possessions as long as they could.

The scene that met one's eyes in the city was fantastic. Civil authority had broken down. The European officials and residents had all evacuated. The white police officers had gone and most of the Indian and Malay constables had returned to their homes in the surrounding villages. There was looting in progress such as I have never seen before. Most of the big foreign department stores had already been whistled clean since the white personnel had gone. There was now a general sack of all shops and premises going on. The milling crowds in the streets were composed chiefly of Tamils, who were the poorest section of the population and therefore perhaps had the greater inducement to loot, but there was also a good sprinkling of Chinese and Malays. The streets were knee-deep in boxes and cardboard cartons and paper. Looters could be seen carrying every imaginable prize away with them. Here was one man with a Singer sewing-machine over his shoulder, there a Chinese carrying a long roll of linoleum tied on to the back of his bicycle, here two Tamils with a great sack of rice suspended from a pole, there a young Tamil struggling along with a great box of the best Norwegian sardines. Radios, rolls of cloth, tins of preserved foods, furniture, telephones, carpets, golf-clubs, there was every conceivable object being fiercely fought for and taken away. One man had even brought an ox-cart into town and was loading it up in the main street outside Whiteaways. The most striking sight I saw was a young Tamil coolie, naked except for a green loincloth, who had had tremendous luck. He had found a long cylindrical tin, three inches in diameter and a foot long, well wrapped up. What could it contain? Obviously a tin like this could only contain some rare and luxurious Western delicacy. He sat on the kerbstone turning the tin round in his hands. He wished that he could read that Western language so that he might know what the tin contained. Should he open it now or should he wait until he got home? Curiosity got the better of him and he decided to open the tin. Carefully he peeled off the paper and took off the lid. Three white Slazenger tennis-balls rolled slowly out, bounced on the pavement and then trickled into the gutter where they soon lost their speckless whiteness. Slowly an expression of the profoundest disappointment spread itself over the face of the young Tamil. He looked at the tin again, and then, with a gesture of supreme disgust, threw that too into the gutter. After a moment's thought he bestirred himself and moved off to see if he could find something more useful—a large roll of red velvet, perhaps, such as was even now being loaded on to a bullock-cart.

The only thing that was not being looted was booze. Several days previously the army had collected as much of the liquor in Kuala Lumpur as it could find. Tens of thousands of bottles and cases were amassed. When the time came for a move south, Local Defence Volunteers laid into the cases of gin and whisky and other intoxicants with sledge-hammers and destroyed them. It was a wise precaution.

We went up to the Residency to see if the Resident was still there. It was a large spacious white house in park-like grounds filled with flowering trees, surrounded at a distance by other official residences. The place was deserted. The flag was down. There seemed to be no-one within miles. The big house was empty. It reminded me somehow of the *Marie Céleste*, that ship which was found in the South Atlantic sailing under full sail but without anyone on board and nothing to show what had happened. In the Residency a half-finished whisky-and-soda stood on the small table by the sofa in the drawing-room. Upstairs a woman's dress, half-ironed, lay on the ironing-table in one of the bedrooms. Two dispatches addressed to the Governor, typed out but unsigned, lay on the desk upstairs. In the offices on the ground floor the files were intact. The staff appeared to have downed pens in the middle of whatever they were doing and made off. A lorry, still in good order, was parked at the side of the building. Cases of beautiful silver ornaments, daggers of superb native workmanship, the presentations, doubtless, of Malay princes, lay in glass cases in the hall. The official portraits of the King and Queen smiled down from the walls....

Those beautiful houses on the outskirts of Kuala Lumpur, those spacious mansions, with their lovely tropical gardens, where bougainvillea and canna and hibiscus and many other flowering shrubs and creepers were in full bloom, were absolutely deserted, save perhaps for an old Chinese servant on the back premises or some dog whose master had not been able to take him south. At the hospital the Indian medical officer told us that all the European patients and staff had left for the south. He was in charge. The Majestic Hotel, which had remained open for so long, thanks to the courage of the Chinese manager, one of those little men who in a crisis reveal unsuspected capacities for courage and strength, was closed at last. Indian sappers were preparing to demolish some of the buildings and sidings at the railway station.

Ian Morrison, *Malayan Postscript*, Faber and Faber, London, 1942, pp. 112–16.

29
Kuala Lumpur Hotels

JAMES KIRKUP AND OTHERS

The first of these passages, from a letter to the Selangor Journal, *expresses the need felt in the 1890s, as Kuala Lumpur first began to attract short-term visitors, for a decent hotel. The first effective response was the Station Hotel, part of the main Railway Station (still standing), designed in 1900 but not completed until 1911. Then, in the ups and downs of the Malayan economy, there was no substantial further progress until 1932, when the Majestic Hotel, here described by James Kirkup in 1961, opened its doors. It was for a generation the scene of many important social and sometimes other gatherings. But the rapid growth of the Kuala Lumpur bureaucracy and business community in the immediate post-war years created an accommodation crisis—and some curious responses to it, as described in the second passage. Kirkup says that he much preferred the Majestic Hotel to 'the swish modern air-conditioned hotels' which had sprung up before his arrival. It is a matter of taste; certain it is that the Station and Majestic Hotels, until the 1950s, offered more civilized lodging to the temporary visitor than the rather down-at-heel lesser hotels which were the only alternative,*

1895

AT present when any visitors arrive—officers of one of the gun-boats, globe-trotters or merchants—they have to be billetted on to various good-natured and hospitable householders, or be consigned (and this is a very bad compliment) to the Rest House. When these visits were few and far between it was all right, and we were all very glad to do our part of the entertainment for which Selangor has such a well-deserved reputation.

How different if we had a good hotel. Many of our guests of an independent turn of mind would far rather stay there, and would, I am positive, come much more frequently if they were not obliged to be dependent on the hospitality of others.

What does your rich globe-trotter say when told of Selangor

as a desirable place to visit? 'Is there a good hotel there?' 'No, none at all.' 'Then I won't go to such a one-horse place!'—and small blame to him.

Men from the out-stations would come in much oftener if they had somewhere to go, more especially married men, who think it rather a tax on their Kuala Lumpur friends to be continually bringing in their family from Saturday till Monday.

Then when, as is so often the case, an unfortunate Government Officer arrives and finds absolutely no provision made for his accommodation by the paternal Government to which he has the honour to belong, he would not feel nearly so hardly used had he a decent hotel to go to.

How invaluable a good hotel manager would be as a caterer for lunches, dinners, suppers at dances and other entertainments.

The bachelor tired of solitude and anxious for a change of diet would find all his requirements at the *table d'hôte.*

1948

The two hotels in Kuala Lumpur at this time were the Station and Majestic Hotels. Both were full to bursting. In 1948 I was put into Bok House, a 'millionaire's folly'. Chua Cheng Bok, in his day one of the wealthiest Chinese in K.L., had built himself this mansion for prestige (with a smaller and more comfortable house behind it for actual occupation by his family). In 1948 an enterprising local character (bandmaster of the FMS Volunteer Force before the war and originally, I believe, a White Russian from Shanghai) had persuaded the Bok family to rent him the mansion to relieve the shortage of accommodation.

The house had a pillared portico. Around the spacious central hall was an array of statues which could be rotated on their pedestals (on wilder occasions they were turned to face the wall). Along one side of the ground floor was a vast reception room with heavy leather armchairs and oil paintings on the wall—rather in the style of a members' reading-room in one of the Pall Mall Clubs—and about the same size. On the other side was a billiard room. Above there was a spacious central landing and on either side a bedroom suite. Each suite contained a vast bedroom, a dressing-room and a bathroom.

VISITORS TO THE F.M.S.

Should stay at the

Station Hotels at IPOH and KUALA LUMPUR.

First Class Accommodation. Excellent Cuisine.
Hairdresser on the Premises. Electric Light, Fans & Lift.

THE LOUNGE, KUALA LUMPUR STATION HOTEL.

THE STATION HOTEL, IPOH.

TARIFF (exclusive of beer, wines and spirits).

Board residence, per day	$9.00	each person.
Bedroom and attendance only, per night	5.25	
Breakfast	$1.00	Served in Restaurant on ground floor.
Tiffin...	1.25	
Dinner	1.75	

Telegrams requesting bedrooms to be reserved for passengers travelling by train will be sent from any Railway Station free of charge.

First-class accommodation for visitors to the FMS, from D. J. M. Tate (Compiler), *Straits Affairs: The Malay World and Singapore*, John Nicholson Ltd., Hong Kong, 1989.

The problem which confronted my landlord (and his wife) was how to squeeze enough people into these huge rooms to make the arrangement a paying proposition; Government officials were considered fair game since it was known that they recovered their rent from public funds. So the more fortunate government officers were accommodated in the spacious but rather uncomfortable bedrooms. As a newcomer I was given one of the dressing-rooms. As my room had the only access to the bathroom and lavatory allocated to male inmates I was curtained off (somewhat inadequately) from the traffic through my room at night—on Saturday nights in particular fellow guests returning from their potations were a nuisance.

But I did better than some. The more impecunious guests were either crowded in together—I remember that at least five women secretaries occupied one of the bedrooms or (at lowest rents of all) one could camp on one of the verandahs, trusting to the bamboo sunshades for privacy from observation from outside. These guests, typically optimists trying to establish one-man businesses in the teeth of Chinese competition, suffered particularly if there was heavy, driving rain blown in on their side of the house. But we all mucked in together quite amicably.

1961

There are a number of swish modern air-conditioned hotels with their own shopping-arcades, restaurants, bars and cabarets. Their pretentious luxury always bored me to extinction. I much preferred the very moderately-priced and charmingly old-fashioned Majestic Hotel, where I stayed for my first six months in Malaya. At first I had an immense double room with two beds cocooned in white oblong mosquito nets ... it was comfortably furnished with massive Edwardian coat-racks, wardrobes, chairs and tables and desks. I also had a sitting-room and a tiled bathroom. My apartment was painted in white and lime-green.... The rooms were all swinging doors, jalousies, vasistas [half doors] and slogging three-bladed metal ceiling-fans which kept sweeping up all my papers into swirling snowstorms.... The oblong mosquito nets looked oddly functional: the boy used to tuck them in well under the mattress and in one side of the net was the sort of aperture which is found in Y-front briefs. I had to crawl through this in order to get into bed.

The bed was delicious in many ways. It had a good firm mattress stuffed with locally-grown fresh kapok and I had a 'Dutch wife', a long bolster for resting the legs on so that the air under the sheets had a chance to circulate and prevent night sweats, and day sweats too for that matter. The pillows were firm, and from them, as from the cupboards in the room, came a curious musky perfume—a mixture of dried poppy and verbena, a touch of mignonette, and something else quite primeval, like dried human hair: the smell is musky-sharp and sleep-inducing. I found the same smell in many other Chinese hotels.

1894: 'A Suggestion', letter to the *Selangor Journal*, 3, 28 December 1894, p. 132, signed 'A Supporter'; *1948*: J. M. Gullick, 'My Time in Malaya', unpublished; *1961:* James Kirkup, *Tropic Temper—A Memoir of Malaya*, Collins, London, 1963, pp. 32–3.

Penang and Malacca

30
Penang in 1867

ARCHIBALD ANSON

The high hopes of Penang's future in the early years of the Settlement were disappointed, leading to a long period of neglect. When, in 1867, responsibility for the Straits Settlements was transferred from the Government of India to the Colonial Office, Colonel (later Major General) Archibald Anson, who had served with the army in India for many years, was appointed Lieutenant Governor of Penang, that is, the resident chief administrator there, and also the deputy of the Governor of the Straits Settlements in Singapore.

In his memoirs Anson describes the 'feeling of depression' which he felt (and why) on taking up his appointment. It was somewhat relieved by having the use (except when the Governor visited Penang) of the residence on Penang Hill, later known as 'Bel Retiro'. The secret society riots of 1867, also the product of long neglect of local conditions, were among the most serious disturbances of their kind in the history of the Straits Settlements.

THE day after my arrival in Penang, I went to the Government Offices. I found these to be in a long narrow one-storied building, with a pandal, or verandah, projecting from below the eaves, and supported by wooden posts. This pandal extended the whole length of one side of the building, and also along the front of it in the main street, over what was the front of the Treasury Office. An open ditch ran under both the side and end of this pandal, up and down which the tide flowed and ebbed. When, some time after, I had this ditch, which was the full width of the verandah, about 5 feet, filled up and converted into a cemented pathway, it was looked upon as a dangerous innovation, that would be sure to cause some mischief.

I imagine the ditch along the building had been made originally to receive the rain water from the roof, before the pandal was added. The tide caused the lower part of the building to be damp. The ground floor of the offices contained a number of stone-paved stores, with iron-grated windows looking out under the pandal; and behind them was a wide, open stone-flagged passage way, with the doors of the stores opening into it. These stores, in the old days of the Indian Government, when, as a Company, trade was carried on, were used for storing anchors, chains and all heavy goods. One store, at the entrance to the building, had been converted into the Treasury Office, and another at the furthest end, close to the sea, had been converted into a Post Office. The first store on the left of the entrance, and opposite the Treasury, was the guard-room for the Treasury guard, from the Madras sepoy regiment stationed at the sepoy lines in the country, about two miles away.

My first feeling on entering the building was one of depression. I had to step down a deep stone step into a gloomy stone-paved passage, where the guard turned out to receive me. The passage was so low that some of the bayonets of the guard punctured the ceiling when their muskets were being brought to 'the present'.

When I proceeded up the staircase, opposite the entrance, to the first floor, I found a broad landing with a suite of three rooms on the left, which were the offices of the Governor of the colony and his staff, when they visited the Settlement. On the right was my office, which opened again into a suite of offices over the stores below. The first of these offices was occupied by my clerks, and then came the Land Office, the Survey Office, the Office of Imports and Exports, and the Harbour Master's Office. The Harbour Master was also Postmaster, and had the Post Office below his office. Along the back of these offices, and over the open passage way below, ran an empty corridor. This corridor and the passage below were used in the days of the old 'John Company' (as the old

Indian Government under the company of directors was called) for the accommodation of those who came to purchase the wares of the company. The soft and lighter goods were stored in these upper rooms....

Penang, on my arrival, appeared to me a very forsaken place, and the customs of its inhabitants very antiquated; and the individuals forming the society of the place expressed their determination to adhere to their old customs, and not to admit any change in them.

George Town, the capital, was in a very dirty and neglected state. The main street had never been macadamized, except in very small patches here and there, and the shopkeepers, mostly Chinese, occupied a considerable portion of each side of the street with their wares; besides blocking up a five-foot path under cover of the front of their houses, which by law was intended for the use of the public.

There had been a bad feeling fermenting between two rival Chinese secret societies for some time before I arrived in the Settlement, and this feeling had increased during the last Mohurrum [Muslim New Year] festival, and led to constant assaults by individuals of the one party on those of the other. This had culminated in the murder of a Malay diamond merchant, a member of one of these societies. There were two societies among the Malays of the Settlement, named the Red Flag Society and the White Flag Society. These societies were originally of a religious character; but that character very soon ended, and they took to quarrelling and fighting with one another. The Red Flag Malays joined the Chinese Toh-Peh-Kong Society, and the White Flag Malays the Ghee Hin Secret Society.

About the beginning of July, a Toh-Peh-Kong Chinaman was looking through the palings bounding the premises of a White Flag Malay, when the Malay threw a rambutan (a Malay fruit) skin at the Chinaman, and called him a thief. The Chinaman went away, but returned with ten or twelve Toh-Peh-Kong friends. The Malay's friends then turned out, and a fight with stones and clubs took place. The Malays drove back the Chinese as far as their kongsee house, the meeting place or club of their hoey. A kongsee is a company; a hoey, a secret society. The stones thrown by the Malays then struck the Toh-Peh-Kong signboard, upon which the Toh-Peh-Kongs turned out in great number, and firearms were resorted to. The police interfered, and, for the moment, stopped the disturbance. After this, frequent assaults and murders were committed by both societies, and on the 1st August a false charge was made by the head man of the Toh-Peh-Kongs, that some White Flag Malays and Ghee Hins had stolen some cloth that, after being dyed, had been put out in the street to dry, by some Toh-Peh-Kong dyers. There is no doubt that this charge was made to bring about a 'casus belli', for which the head man had made every preparation. On the 3rd August the Toh-Peh-Kongs attacked the Ghee Hins, and thus commenced the great Riot of 1867....

The Government House, or Government bungalow as it was called, on the mountain, consisted of two bungalows, joined by a corridor, 168 feet long, from which there was a magnificent and extended view, north and south, for a distance of about 90 miles of sea, with islands dotted about. The view to the west was across the strait dividing the island from the peninsula opposite; and then across the peninsula to the mountains forming its backbone. To the east was the jungle, which came close up to the bungalow on that side; and in which monkeys frequently sat on the trees, within sight of us, when we were playing billiards in the corridor.

I found the garden and grounds in a very neglected state, but, with the aid of the Indian convicts employed there for the upkeep of the grounds and the road up from the plain, made great improvements. There was in the garden, opposite one end of the bungalow, a monument. This had been erected, by the aide-de-camp of a former Indian Governor, in memory of a dog that had belonged to the Governor's daughter. It appeared that the officer was engaged to the young lady, and sent to Calcutta for the stone urn which was on the top of it. The pillar on which the urn rested was square, and about 3 feet high, and had a poetical inscription, covered with glass, let into the front of it. The engagement between the young lady and the officer was broken off, and then he sent her the bill for the urn.

There was a signal station close to the bungalow, at one end of it, which signalled the arrival of vessels, and, by means of an old-fashioned telegraph worked by shutters (afterwards replaced by an electric telegraph), communicated with Fort Cornwallis, on the seashore, at the town. This telegraph, and the convicts, were in charge of an English pensioner from the

Madras artillery. His name was O'Neill; and when I was at Singapore, I one day received a telegraphic message which, being badly copied, purported to come from 'One ill', and I was greatly puzzled for some time to make out from whom it was sent.

A. E. H. Anson, *About Others and Myself 1745–1920*, John Murray, London, 1920, pp. 275–9, 289–90.

31
Putting In at Penang in 1906

JAMES ABRAHAM

Although it was overshadowed by Singapore as the political and commercial capital of the Straits Settlements, Penang thrived as a centre of trade with northern Malaya and Sumatra, Burma and southern Thailand. Travellers to Malaya from the west, whether they were visitors or arriving to pursue their careers, disembarked at Penang unless their final destination was Singapore or southern Malaya. The railway terminal was on the mainland opposite Penang. Hence, Penang was a busy port, as here described by the surgeon of a large steamer, which had put in to unload part of its cargo, in the course of a voyage to Japan. The streets were 'gaily decorated' in preparation for a visit by the Duke of Connaught, which dates this account to February 1906, although it was not published until 1911.

THE first impression was of a wonderful green: the land seemed smothered in vegetation. It rose precipitous from the water's edge, crag upon crag of naked rock jutting out grey amongst the green, with here and there the white outlines of verandahed bungalows, perched perilously on the heights, which, half hidden in the verdure, rose higher and higher, and culminated finally in one great peak 2,700 feet above the sea....

We anchored opposite the jetty in thirteen fathoms of water; and the first thing that struck us was that it had suddenly become intensely hot—we were no longer making a breeze for ourselves. The next impression was that we were being boarded by pirates. They came from every side, sampan racing sampan for which would be the first to reach the lowered gangway. They tumbled on deck in heaps from every quarter. In five minutes they had penetrated to every corner of the ship—Parsees, Malays, Klings, Chinamen, and Eurasians. There were money-changers with bags, clinking the large silver Straits Settlements dollar, cigar merchants selling Burmah cheroots, tailors wanting to measure one for white suits, men with fruit of tropical lusciousness, boys with the inevitable picture-postcards.

Almost before the engines had stopped a series of lighters, with great bamboo masts and yards, began to arrange themselves around the ship. Scores of brown, half-naked, turbaned coolies swarmed on board, opened the hatches, and with naked feet on the levers started the steam winches running. In an almost incredibly short period after our arrival cargo was going over the side into the empty lighters, and khaki-clad Chinese tally-clerks in puggarees, standing one at each hatch, were checking the loads as they rose from the hold.

The heat was sweltering. Every one was busy—the officers looking after the cargo, the 'Old Man' closeted with the agent, the Chief [Engineer] seeing about the supply of fresh water which was being pumped from the lighters into the tanks.

The Chief Steward was going ashore to order fresh provisions; so we took a sampan together.... When we got to the bottom of the landing stage we saw two long rows of 'rickishaws, one on either side. The owners of the nearest two leapt across the road, whirled their light vehicles round, and stood grinning till we mounted....

We sped along wide open streets, lined by Chinese shops, past patient oxen dragging springless carts, past itinerant merchants carrying their stock-in-trade in large hemispherical baskets, slung, one on either end of a long bamboo pole, over one shoulder, and held by the corresponding arm, while the unoccupied hand worked a wooden rattle to attract the attention of possible customers, past big Sikh policemen, who gravely saluted when we paused to look at them, past Chinese temples, dragon-haunted, lantern-hung, along a gaily decorated road, past an open space where little pigtailed Chinese boys were playing football, barefooted, with the temperature at

95° F. Other 'rickishaws met us, carrying pale-faced Europeans dressed in white, with white pith helmets like ourselves. They all stared at us. Sometimes a passing 'rickishaw would carry a portly Chinese merchant, or a Chinese woman with death-like, white-chalked face and henna'd lips, or little Chinese girls with tinsel crowns and flowers in their hair—for the celebration of Chinese New Year was not yet over.

James Johnston Abraham, *The Surgeon's Log*, Chapman & Hall, London, 9th edn., 1916, pp. 63–7.

32
The E & O Hotel at Penang in 1928

GEORGE BILAINKIN

The author was a young journalist who arrived in Penang about 1928 to take up a post on a local newspaper, the Straits Echo, *whose editor was Richard Sidney (see Passages 16 and 27 above). However, Sidney was evidently not a success and almost immediately Bilainkin succeeded him.*

The E & O Hotel still stands looking across the straits at Kedah Peak—surely one of the most beautiful 'view of the sea' offered by any hotel.

DAWN came as we awoke in the mirrory waters of Penang, and looking across to the Malay Peninsula, I saw a few clouds in the south moving away gently. The lightening sky was growing redder and redder, as if human hands were holding a palette, and using the brush indiscriminately and with an urgent speed. The sea was still, and then tiny ripples rose as strange craft, worked by Chinese hidden under huge hats, passed in the distance. With the stronger sun the colour of the water became deep blue, but not so deep that it lost transparency.

From the deck I watched the ebullition of spirit among passengers in the second saloon where there was less restraint over joy and sorrow. They babbled and said nothing, but seemed human in their heartfelt welcome on seeing land.

Among the first class passengers the greatest anxiety was to settle up tips with stewards, to ensure the baggage being taken to the right hotel, and to photograph in the mind the dusky men swallowing one's possessions from the cabin.

As I stepped on shore, having told the office messenger that I should see my chief at the hotel at breakfast time, I was at last alone and free to think. It was pleasant to stand on a soft tarmacadam surface. I glanced at the long line of motor cars waiting to take the passengers whither they willed.

Which should I choose? Suddenly I seemed to be a prisoner, surrounded by no fewer than a dozen little carriages, not unlike the bath chairs in the parks of Bath or Kensington Gardens. In front of the place for the passenger, who must sit high up, between two enormous wheels with rubber tyres, were the shafts. Inside these stood Chinese. They wore little blue slips and a pair of dark shorts. Some had dark shorts only, and no hat. One smiled under a torn straw hat. They were shouting 'Runnymede' and 'E. and O.'. The words sounded differently but I knew that was what they must mean. They were jostling and pushing each other out of the way. One man bawled loudly, and, was taller than the rest; he had a strong chest, brown and bare. He came nearest to me. Two elderly men, who seemed to be toothless, and who before the shouting had stood close to me, again receded.

Hitherto I had felt sympathy with ricksha men as a whole, but now my concern was for the weaker old men. I must have remained still for at least two or three minutes, drinking in the scene, without utttering a word. The gabbling continued—broken English words, and Malay intended for me, and Chinese addressed to fellow members of the attack. One man came and almost wrenched one of the two little attaché cases out of my hand. But he promptly retired amid loud laughter from his competitors. I walked slowly through the smell of old shirts and shouting humanity to the oldest coolie who now stood away at the back of the group. The others knew by the uncertain manner of my mounting that I had never been inside a ricksha. They giggled and shouted. I sat back and discovered how comfortable a ricksha could be. The old man almost doubled himself up between the shafts, turned the carriage to the right and to the left, looked back to see whether there was any traffic behind, and was off. The master was on the throne. The slave was in harness.

His shoulders moved in rhythmic action—just like those of a trained marathon runner. I half shut and then opened my eyes and saw the perspiration on the back of the man's vest. It was I who caused this perspiration.

Up the slight incline past a palatial building opposite a cricket ground, the coolie slowed down. He wanted to overtake a ricksha in which two fat Chinese were holding a large wooden box. He shouted 'Eh'. The coolie understood and gave us right of way. The hood was down, but I felt the fierce glare and stifling heat of the sun despite my huge topee. I was sitting in a sort of comfort but the old man in front was running. It seemed a crime to use him thus, and a needless crime.

I stumbled out of the carriage on arriving at the hotel and offered him twenty cents, nearly twice the sum to which he was entitled, twelve cents. The coolie wiped his face with a towel, then the neck, and laughed. My attaché cases were taken from me by a small fat Indian wearing a well preserved cap and uniform with brass buttons, who spoke English. His khaki trousers were incongruous, for I glanced down, and saw that like all labouring Orientals he wore no socks or shoes.

A luggage sticker of the E. & O. Hotel, Penang, *c.*1920, Raffles Hotel Collection.

Chinese clerks in the reception office spoke in excellent English, and I felt surprise to find them wearing the same kind of tropical suit of white as I was, instead of a fantastic Mandarin coat of a style fostered in my imagination by old cinema films.

When I asked them about a table in the dining room, as directed by the knowing folk on board ship, the clerk promised that Hindenburg would see to all that. I repeated the name in slight astonishment and a European behind me explained that the reference was to the head boy in the hotel.

From nowhere arrived a smiling miniature replica of Hindenburg, head closely shaved, of medium height, obvious Mongolian stock, but not strikingly Chinese. He bowed and said, 'Tuan would like a table?' I replied that I should, preferably one in a quiet spot. His conclusions were not to be read. I was aware, however, of his noticing the creased appearance of my white suit, for it had not been ironed successfully by my Irish cabin steward. In the style of an accomplished English butler, Hindenburg led me with another stage bow to a table, immediately below the orchestra! Happily they were not playing at breakfast time.

I thanked Hindenburg as warmly as one would thank a distinguished maître d'hotel elsewhere. He bowed again, smiled and moved away noiselessly. I liked his black silk shoes, white socks, black trousers which reached a few inches below the knee, and a black coat of thin material. It was closed and fitted tightly round the neck. The huge head, the almost stern expression as he spoke to two Chinese boys near the table, made me see instantly his likeness to the German war leader.

The elegance about his walk and dignity were more fascinating to watch than the half-hidden faces of bored Europeans reading the newspaper which I was to help edit in a few days.

I counted at least two dozen 'boys', all dressed alike and walking about noiselessly. The hair of everyone was carefully cut; many had partings, and not a few had apparently adopted the European device of using brilliantine or some other kind of grease. There seemed to be a waiter to every two or three tables. Among the rows walked Hindenburg and another elderly 'boy', helping a guest into a chair, or drawing the chair back when he had finished the meal.

A boy came up to my table, bowed, and showed me the

long menu of meats, hot and cold, several kinds of fish, about a dozen egg dishes, savouries and much else! And for some unfathomable reason I had imagined that in the East—Penang is not yet the Far East, which begins east of Singapore—the food must be 'primitive'....

George Bilainkin, *Hail, Penang! being the Narrative of Comedies and Tragedies in a Tropical Outpost, among Europeans, Chinese, Malays, and Indians*, Sampson Low, Marston & Co., London, 1932, pp. 10–13.

33
The Jewel of the Orient

DONALD MOORE

Donald Moore was an active and innovative publisher in Singapore in the period after the war. He commissioned and published a variety of short studies of contemporary (and also historical) subjects, to which he himself contributed as an observer of the Singapore and Malayan scene.

This entry in Donald Moore's published diary is dated 29 August 1952. Since that time there have been improvements, such as the construction of a bridge across the Penang straits, and provision of improved workers' housing. But much of the passage still aptly describes what may yet be seen in parts of Georgetown and some other Malayan towns.

WHEN the traveller arrives in Penang he may well experience some difficulty in understanding how it came to be described as the 'Jewel of the Orient'. If he arrives by train he will, after the crossing from the mainland in the ferry, be ejected on to the hideously decrepit Malayan Railways pier. This dark and dank excrescence, this gateway to Penang, enshrines the very essence of 'tidapathy' [indifference], a dilapidated commentary on a Railway Administration or a Municipality either too idle or too poor to provide anything less second rate.

The exit from the docks is hardly more impressive. The traveller will find that when he is in Beach Street he is in

Penang's main business quarter, and that when he has turned right into Bishop Street he is in the main shopping area. Both are characterized by meanness, by the oppressive atmosphere of dacaying impermanence where no thought has ever been taken for the morrow. Although buildings of fairer proportions are now dotted among the unlovely shophouses, the need for much destruction and rebuilding remains before the commercial community can claim to carry on its business in an area architecturally more in tune with the profits it has made in the past.

The traveller by air will receive perhaps the most favourable impression, but even he, during the long drive into town, will scarcely be excited by anything spectacularly beautiful in the passing scenery: coconut trees, some rubber, rice, patches of vegetables, and the inevitable 'rubber cows' as the black buffaloes have been aptly called. As he enters Penang town or, to give it its correct title, Georgetown, he must skirt the edges of Chinatown that in no way differ from any other South-East Asian Chinatown—an area which, according to the temperament and intelligence of the visitor, can be exciting, exotic, seeped in the romance of the Orient or simply slums of a most disgraceful nature.

Chinatown consists basically of shops in their crudest form: a ceiling, a floor, three walls and an open front. People live in these shops in great profusion; they live in even greater profusion behind them and above them, and some, the homeless, live in front of them in the covered five-foot ways, separated from the seething streets by the deep monsoon drains that carry away the filth of Chinatown as well as the rains. From every window, from the cubicled blackness behind the drooping, leaning, decaying walls, project poles on which hang the day's washing; to its everlasting credit the Chinese race may live in the grime of centuries and yet remain as clean, both in its person and in its clothing, as any race on earth.

If the visitor is of the more unusual persuasion who finds nothing admirable in the sight of Oriental slums, he will at least be struck by the abounding energy of those who live in them. He will see shops of every description: fish shops, polluting the air for streets around, selling big fish and little fish, live fish and dead fish, wet fish and dried fish, fish caught on the coasts by Malays with primitive nets, fish trapped by Chinese at night, enticed into bamboo cages by the lights of hundreds of gleaming pressure lamps. He will see funeral shops where are sold the joss sticks for the temples, and the paper money thrown away to placate the evil spirits and the incredible mountains of paper finery that adorn the coffins on their journey through the streets to the burial grounds. His ears will be assaulted by the din of blacksmiths' shops and locksmiths' shops, and bicycle repair shops and Chinese gramophone record shops and cabinet makers' shops, and the hammering of nails into picture frames, packing cases, the soles of shoes and bedsteads. His mind will reel before the vast profusion of blood-red Chinese characters that cover every inch of wall space, and his stomach will turn in the countless coffee shops where in stained, cracked cups, is served the worst coffee in the world. And if the visitor looks further, if he goes behind the rotting, dhobi-festooned façade, leaving behind the clamour of ten thousand clack-clack-clacking wooden clogs on concrete, the roar of rain on rusting corrugated iron and the babble of voices that never stills but only ebbs and flows with every passing day and night, he might well come across the horrors that lurk in this cathedral of poverty: women renting spaces under beds to deliver themselves of unwanted babies in the only peace they can find, abortionists blindly probing and prodding with primaeval appliances the women who in desperation turn to them only to die in agony. He will stand helpless before the bodily starvation of opium addicts, young prostitutes not yet in their twenties, drinking caustic soda to end it all, or living to die old women at forty, leaving a dozen unwanted, under-nourished, diseased children, behind them.

And the visitor in the evening may walk up Northam Road, and further, and look out to sea. The little waves will lap quietly at his feet and the palm fronds will whisper in the lightest and freshest of breezes. He will look into the mist of distance, along the dappled white track of the moon, and see nothing more than the bob of the lights of far-away fishing boats and hear nothing more than the high drone of cicadas. On either hand the bright clean sand will stretch to the rocks that form the end of one bay and the beginning of another. Behind him, sprinkled with starry lights, will loom the glorious hill of Penang.

He cannot but think himself in heaven. He will say: 'This is indeed the Jewel of the Orient.'

Donald Moore, *Far Eastern Agent or the Diary of an Eastern Nobody*, Hodder & Stoughton, 1953, pp. 17–20.

34
Malacca in 1880

AMBROSE RATHBORNE

On his first arrival in Singapore (from Ceylon) in 1880 Rathborne (see Passages 17 and 24 above) embarked in a coastal steamer, of which his impressions are very like those of Ethel Douglas Hume (Passage 10 above) almost twenty years later, for Malacca, whence he made his way to begin his Malayan career in Sungei Ujong. Until the railway line from Penang to Johore Bahru, opposite Singapore island, was completed in 1909, travellers to the western Malay States followed this route—unless it was more direct to travel from Penang to the northern States (Passage 31 above).

Malacca was now a minor port, with a long history, but no longer of any commercial importance except as a 'feeder' to and from Singapore by coastal steamer.

AS the steamer which was to take me part of the way [from Singapore] towards Sungei Ujong—at that time a state but little known—started from the outer harbour, I stepped into a sampan from the jetty close by the clubhouse, and was soon alongside of the steamer, where the noise of the donkey-engine, the rattling of chains, the loud quacking of the ducks as they were slung on board, betrayed the activity that was going on in order to get the ship off to time.

The decks were piled with crates of fowls and ducks; jars of spirits, tubs of fish, cases of kerosene oil, lined the sides; and on the hatchways being closed they were immediately taken possession of by the many deck passengers, who opened and spread their mats upon which they reclined, making themselves as comfortable as they could. An awning was stretched across overhead as a protection against sun and rain. Every vacant nook was occupied by some being or some thing, a most varied and miscellaneous assortment as we looked down upon it all from the poop.

The steamer seems to have no buoyancy, for she is below her Plimsoll mark and overloaded, the passengers are greatly in excess of her licensed number; but what matter? Europeans seldom travel in her, she is under the control of a Chinese supercargo who thinks only of profit, and the port rules were not at this time stringently enforced; so as much as could be crammed on board was taken, until the lapping of the water high up along the steamer's sides warns even the most heedless that it were folly to load her more.

The whistle sounds, the rattle of the chain is heard as the anchor rises, the steamer slowly threads her way through the shipping, and turns to go a short way into the Straits of Malacca, by passing through the inner harbour where there is a narrow channel between two islands. On the one side wharves stretch from end to end, alongside of which big ocean-going steamers are moored. The number of warehouses behind the

In this early nineteeth-century view of Malacca, the church is on the left and the stadt house (government building) on the right, from Capt. P. F. Begbie, *The Malayan Peninsula*, Vepery Mission Press, Madras, 1834.

quay testify to the large trade being carried on, for we are just leaving the great emporium of Malaya, where steamers coal on their outward and homeward voyages, and where goods are transhipped and distributed to all parts of the world. The docks are situated here; ships are being repaired, and the constant clang of hammers is heard as we [pass] along.

On the other side a silent dulness pervades the reddish-coloured hills, the summits of which have since been planted with trees, giving cover to strong fortifications and big guns, which it is hoped make Singapore impregnable to attack from the seaward, whilst boards at the water's edge mark the position of some submarine mines placed across the narrow entrance to the harbour. Passing between two red bluffs, we emerge into a well-buoyed channel that guides us into the straits beyond.

The sun has set, the outline of the coast fades from view, the night is fine and clear, a cool breeze makes a ripple on the water, and tempts me to have my bed made up on deck; but sleep has hardly come when the noise of hurrying feet is heard, then the flapping of the side awnings as the sailors lower them and make them fast: none too soon, for a black bank of clouds has formed ahead, the breeze has freshened, and we are soon struck by the squall, accompanied by torrents of rain and blinding darkness. The steamer slacks her speed, begins to pitch, and the waves splash over her dipping bows; her human freight seeks shelter as best it may. An hour, and all is over; dripping decks, dead and dying ducks and fowls that have been smothered or trampled upon by their fellows, and small crested waves alone remain to tell of the squall we have just gone through, an experience very frequent in these waters at certain seasons of the year.

Early the next morning we approach Malacca, and on nearing the coast perceive that the whole shore is lined with a deep fringe of cocoanut-trees, whilst in the distance can be discerned the hills and mountains which form the dividing range of the Peninsula. We pass a number of small canoe-shaped fishing boats, in each of which a man is seated, having a wide conical hat upon his head, which looks like an inverted mushroom, as a protection from the sun. The tiny boats rock to every ripple, causing anyone affected with seasickness to shudder as he watches their occupants quietly fishing, entirely undisturbed by the pitching and tossing of their little craft. As we draw near the anchorage speed is slackened, but before the engines are stopped the ship is waylaid by a number of boats, which the rowers fasten on to her sides and up which they clamber, and rush amongst the passengers to seek for fares. They vociferate excitedly, for competion is keen, and pick up and secure the goods of their bespoken fares to prevent their changing their minds and engaging their passage to the shore with someone else. These boats are rowed by four or six men; they are capacious and strong, carrying a large sail, and able to withstand rough weather, a necessary attribute, as owing to the shallow foreshore steamers are obliged to anchor more than a mile out, and the journey to and fro is sometimes unpleasantly lively. . . .

The outskirts of Malacca are thickly populated by Malays, whose gardens, bordering on fine stretches of paddy fields, contain both fruit and cocoanut trees, in whose produce there is quite a brisk trade at certain seasons of the year; and the decks of the steamers trading to Singapore are crowded with a miscellaneous assortment, amongst which are the far-famed and prickly durian, so obnoxious to new comers, but delicious to those who have succeeded in overcoming its peculiar smell and taste; dukas, a fruit for which Malacca is famous, and equal in flavour to a nectarine; mangosteens, encased in a peculiar covering that hardens as the fruit ripens, and has a little crown on the top with a leaf for every division of the fruit inside, which is delicate in flavour, but does not keep good for many days; wholesome langsats that have a pleasant acid taste, but whose stone is bitter; red rambutans, with their prickly skins, beneath which a large seed is thinly coated with a luscious covering; and the rambai, a fruit that hangs down in bunches like yellow grapes, full of seeds inside, and somewhat bitter in its flavour. Besides these there are the tampuni, pulasan, papaya and guava, all edible and pleasant to the taste.

The red colour of the roads is an agreeable change from the white granite thoroughfares of Singapore. Good laterite is easily procurable, and the streets are paved with this material, which is softer than granite, and pleasanter to drive along in fine weather.

The sea-wall is faced with blocks of hard laterite, having a honeycombed appearance on the outside. These stones are quarried and cut in a soft state, and after exposure to the air and weather for a year they become hard, making excellent building material, and are much used. The esplanade is lined with flamboyant trees, or 'flame of the forest' as it is sometimes called, and when in blossom the flowers form a mass of gorgeous colour.

No description of Malacca would be complete without mention being made of the Malacca Babas, who are Malacca-born Chinamen, and form a considerable community, many of them being ignorant of the language of their forefathers, and only speaking Malay. They are of a gregarious disposition, and even the wealthy live in their business houses in the town, although they have fine residences and gardens some little distance away in the country, to which they only resort for recreation and change.

The interior of Malacca is less thickly populated, the country is undulating, and the principal settlements are within easy reach of the roads, which are numerous. Wherever there are valleys and gulleys the opportunity has been taken to convert them into paddy fields, which once a year are prepared for planting by having the weeds and grasses growing in them cut down and burnt. Then they are inundated with water, the ground is dug with a long-handled hoe, buffaloes are driven to and fro to churn up the earth, or a small wooden plough is used, depending on the nature of the soil; the water is then let off, the surface smoothed and made ready to receive the plants.

Ambrose B. Rathborne, *Camping and Tramping in Malaya: Fifteen Years' Pioneering in the Native States of the Malay Peninsula*, Swan Sonnenschein & Co., London, 1898, pp. 17–21, 25–6.

The Malay States

35
The Royal Court of Perak in 1826

JAMES LOW

By treaties made in the mid-1820s Britain, Holland, and Siam established an ill-defined partition of their spheres of influence in maritime South-East Asia. The Dutch conceded to Britain an exclusive interest in the Malay Peninsula (of which only Malacca and Province Wellesley were under any form of British control) and Siam, with more reluctance, agreed not to interfere any more in the independent State of Perak (to which they had previously sent envoys and troops) unless invited to do so.

James Low, a British administrator, was sent to Perak in 1826 to stiffen the resolve of the Sultan to oust his unwelcome Siamese guests and assert his authority over his own subjects, in particular those of them who were the 'pro-Siamese' party. Low's account of his mission reflects these tensions, especially in the description of the court ceremony at the end of this passage. But this passage is also of more general interest as a picture of the lifestyle of a Malay State at a time before it was much affected by Western influences.

WE then came in sight of Allaham,* the temporary residence of the Raja who is entitled Sri Sultan Abdullah Ma Allam Shaw. On approaching the landing place

*Allahan (here misspelt Allaham) is on the east bank of the Perak River upstream from its junction with the Kinta River; a little further up-river (on

the Raja saluted us with fifteen guns—and on landing we were met by him in person with his cavalcade of sword bearers and spearmen who conducted us to his house.

The Balei or Hall of Audience was found to be in the usual rude Malayan fashion. It is about fifty feet long by thirty broad and raised in the middle for a space of fifteen yards broad along the length. On this a sofa and chairs were placed and the floor was covered with a Turkey patterned woollen carpet. The Raja's people manifested the most lively curiosity respecting us, having with the exception of those accustomed to trade to Penang rarely seen a European.

The wife of the Raja and her female attendants were observed peeping through the apertures made in the partition at the extremity of the room. The Raja treated us with the utmost civility and we then left him in order to arrange the distribution of the guard and followers and get ourselves and them housed....

The Raja is a very quiet person and very indulgent to his subjects. The routine of his life admits not of much variety. He hears complaints and settles business early in the morning, breakfasts about 10, and dines about sun-set or later.

The people who inhabit the interior have simple and rather pleasing manners. They are strict Mussulmans [Muslims], saying prayers three or four times a day and on weekly fast days the Raja assembles all his officers and attendants in his Mosque and repeats along with his Imams and them select passages from the Koran, with a fervour which is rendered ludicrous by the odd gestures by which it is accompanied. As they become warmed by their devotion they nod and shake their heads violently in concert. One who has not been accustomed to such an exercise would be rendered quite giddy by it. They did not seem the least displeased at my closely observing them. On leaving the mosque every one goes about his common occupation.

One day during my stay the Raja gave a feast to his people, it being an anniversary. A long and slightly built shed was prepared as a kitchen. Here five or six huge iron pots were placed over fires. In each of these about thirty fowls were boiled.

---

the west bank) is Pasir Salak where the first British Resident, J. W. W. Birch, was killed in November 1875. Each Sultan of Perak sited his royal capital at a different place from his predecessors.

They live however very plainly, fish, rice and a little seasoning with fruits, being their common food. Very few of them will taste wine, and none ardent spirits.

The Raja and his people dress themselves in very ordinary garbs, except on occasions of ceremony and at these periods they are only clothed from the waist upwards, besides having on the usual nether garments....

The Raja and his chief men have floating houses in which they can remove to a distance. The raft or boat of the former is known by two poles held upright by persons appointed to the duty the one in the prow the other at the stern.

The Raja accepted on two or three occasions my invitation to tea—when he partook of coffee, biscuit and confections. We amused him with music particularly that of the violin, Captain Elwon being a superior performer as an amateur. One or two of his relations shewed themselves tolerable proficients, but although by no means deficient in ear they appear to have inferior skill to the Malays of Malacca and other Eastern Settlements.

The airs they played had considerable melody but were on a low key and uniformly plaintive. They were excessively delighted however with the brisk airs of Europe. It would be superfluous here to describe Malayan instruments. The large gong at the Raja's house, which was struck at intervals during the night, harmonized by its deep and solemn tone with the sequestered situation in which it was heard.

The Raja having requested my presence at the ceremony of administering the oath of allegiance to some ministers and officers, I accordingly attended at the hall. A large concourse of people were assembled. The chiefs and their attendants were seated on carpets and mats on the floor. In front of the sopha on which the Raja sat, were arranged the following articles, a low stool on which lay the Koran, and a large jar of consecrated water, on the top of which was a model of a crown. The Raja advancing dipped the regalia, consisting of armour, in the water, and placed them against a pillow.

The new ministers and other officers then approached and had the oath tendered to them. This oath consists of two parts and is very short. The first part is the promise of fidelity, the second imprecates every calamity to afflict the juror and his family to remote generations should he betray the trust and confidence reposed in him. The characteristic levity of the Malayan disposition was not even repressed by this solemn act, for the Raja and some of the chiefs indulged their mirth occasionally, to the evident mortification of some of his chiefs then present whose gravity was ludicrously contrasted with it. Several of those who had intrigued with the Siamese betrayed evident symptoms of alarm. Indeed under a less indulgent Prince they must have lost their heads. The ceremony was concluded by a discharge of fifteen guns—after this we partook of some preserved fruits and confections composed of rice, flour and sugar and having shaken hands with the Raja and his principal men, a custom they adopt most heartily, we returned to our temporary home. The Raja presented to me the Kris which he had hitherto worn and a handsome spear—and received in return two plated table candlesticks and shades with some silver plate and inferior articles.

Having staid nearly a month in the country, and the object of the mission having been effected, we left Allahan and returned to the vessel. The simplicity, frankness, and social disposition of the people made an impression upon my mind which cannot be speedily effaced.

James Low, 'Observations on Perak', *Journal of the Indian Archipelago,* Vol. 4, 1850, pp. 500–4.

36
Kuala Kangsar in 1879

ISABELLA BIRD

From Perak early in the nineteenth century (Passage 35) we move on to the first years of the colonial period when the architect of the Anglo-Malay partnership, the talented but eccentric Hugh Low, was Resident (1877–89). Isabella Bird, successfully completing her journey by elephant (Passage 1 above), arrived at a time when Low was away on tour. The reception accorded to her was hospitable but quite extraordinary. The latter part of this passage conveys the rapport which was immediately established between Hugh Low and his no less exceptional visitor. The siamang *to which she refers was a third ape, an untrained gibbon which lived on the roof of the Residency. That building, which had been the house of a Malay aristocrat, was little changed by its current owner who 'grudged every dollar spent superfluously'.*

I was received by a magnificent Oriental butler, and after I had had a delicious bath, dinner, or what Assam was pleased to call breakfast, was 'served'. The word 'served' was strictly applicable, for linen, china, crystal, flowers, cooking, were all alike exquisite. Assam, the Madrassee, is handsomer and statelier than Babu at Malacca; a smart Malay lad helps him, and a Chinaman sits on the steps and pulls the punkah. All things were harmonious, the glorious coco-palms, the bright green slopes, the sunset gold on the lake-like river, the ranges of forest-covered mountains etherealising in the purple light, the swarthy faces and scarlet uniforms of the Sikh guard, and rich and luscious odours, floated in on balmy airs, glories of the burning tropics, untellable and incommunicable!

My valise had not arrived, and I had been obliged to re-dress myself in my mud-splashed tweed dress, therefore I was much annoyed to find the table set for three, and I hung about unwillingly in the verandah, fully expecting two Government clerks in faultless evening dress to appear, and I was vexed to think that my dream of solitude was not to be realised, when Assam most emphatically assured me that the meal was 'served', and I sat down, much mystified, at the well-appointed table, when he led in a large ape, and the Malay servant brought in a small one, and a Sikh brought in a large retriever and tied him to my chair! This was all done with the most profound solemnity. The circle being then complete, dinner proceeded with great stateliness. The apes had their curry, chutney, pine-apple, eggs, and bananas on porcelain plates, and so had I. The chief difference was that, whereas I waited to be helped, the big ape was impolite enough occasionally to snatch something from a dish as the butler passed round the table, and that the small one before very long migrated from his chair to the table, and sitting by my plate, helped himself daintily from it. What a grotesque dinner party! My 'next of kin' were so reasonably silent; they required no conversational efforts; they were most interesting companions. 'Silence is golden', I felt; shall I ever enjoy a dinner party so much again?

My acquaintance with these fellow-creatures was made just after I arrived. I saw the two being tied by long ropes to the verandah rail above the porch, and not liking their looks, went as far from them as I could to write to you. The big one is perhaps four feet high and very strong,and the little one is about twenty inches high. After a time I heard a cry of distress, and saw the big one, whose name is Mahmoud, was frightening Eblis, the small one. Eblis ran away, but Mahmoud having got the rope in his hands, pulled it with a jerk each time Eblis got to the length of his tether, and beat him with the slack of it. I went as near them as I dared, hoping to rescue the little creature, and he tried to come to me, but was always jerked back, the face of Mahmoud showing evil triumph each time. At last Mahmoud snatched up a stout Malacca cane, and dragging Eblis near him, beat him unmercifully, the cries of the little semi-human creature being most pathetic. I vainly tried to get the Sikh sentry to interfere; perhaps it would have been a breach of discipline if he had left his post, but at the moment I should have been glad if he had run Mahmoud through with a bayonet. Failing this, and the case being clearly one of murderous assault, I rushed at the rope which tied Eblis to the verandah and cut it through, which so startled the big fellow that he let him go, and Eblis, beaten I fear to a jelly, jumped upon my shoulder and flung his arms round my throat with a grip of terror; mine, I admit, being scarcely less.

I carried him to the easy-chair at the other end of the verandah, and he lay down confidingly on my arm, looking up with a bewitching, pathetic face, and murmuring sweetly '*ouf! ouf!*' He has scarcely left me since, except to go out to sleep on the *attap* roof. He is the most lovable, infatuating, little semi-human creature, so altogether fascinating that I could waste the whole day in watching him. As I write, he sometimes sits on the table by me watching me attentively, or takes a pen, dips it in the ink, and scribbles on a sheet of paper. Occasionally he turns over the leaves of a book; once he took Mr Low's official correspondence, envelope by envelope out of the rack, opened each, took out the letters and held them as if reading, but always replaced them. Then he becomes companionable, and gently taking my pen from my hand, puts it aside and lays his dainty hand in mine, and sometimes he lies in my lap as I write, with one long arm round my throat, and the small, antique, pathetic face is occasionally laid softly against mine, uttering the monosyllable '*ouf! ouf!*' which is capable of a variation of tone and meaning truly extraordinary. Mahmoud is sufficiently polite, but shows no sign of friendliness, I am glad to say. As I bore Eblis out of reach of his clutches he threw the cane either at him or me, and then began to dance.

That first night tigers came very near the house, roaring discontentedly. At 4 A.M. I was awoke by a loud noise, and looking out saw a wonderful scene. The superb plumes of the coco-nut trees were motionless against a sky blazing with stars. Four large elephants, part of the regalia of the deposed Sultan, one of them the Royal Elephant, a beast of prodigious size, were standing at the door, looking majestic; mahouts were flitting about with torches; Sikhs, whose great stature was exaggerated by the fitful light,—some in their undress white robes, and others in scarlet uniforms and blue turbans—were grouped as onlookers, the torch-light glinted on peripatetic bayonets, and the greenish, undulating lamps of countless fireflies moved gently in the shadow.

I have now been for three nights the sole inhabitant of this bungalow.... Three days of solitude, meals in the company of apes, elephant excursions, wandering about alone, and free, open air, tropical life in the midst of all luxuries and comforts, have been very enchanting. At night, when the servants had retired to their quarters and the apes to the roof, and I was absolutely alone in the bungalow, the silent Oriental sentries motionless below the verandah counting for nothing, and without a single door or window to give one the feeling of restraint, I had some of the 'I'm monarch of all I survey' feeling; and when drum beat and bugle blast, and the turning out of the Sikh guard, indicated that the Resident was in sight, I felt a little reluctant to relinquish the society of animals, and my 'solitary reign', which seemed almost 'ancient' also.

When Mr Low, unattended as he always is, reached the foot of the stairs the retriever leapt down with one bound, and through the air over his head fled Mahmoud and Eblis, uttering piercing cries, the siamang, though keeping at a distance, adding to the jubilations, and for several minutes I saw nothing of my host, for these creatures, making every intelligent demonstration of delight, were hanging round him with their long arms; the retriever, nearly wild with joy, but frantically jealous; all the creatures welcoming him more warmly than most people would welcome their relations after a long absence. Can it be wondered at that people like the society of these simple, loving, unsophisticated beings?

Mr Low's arrival has inflicted a severe mortification on me, for Eblis, who has been absolutely devoted to me since I rescued him from Mahmoud, has entirely deserted me, takes no notice of me, and seems anxious to disclaim our previous acquaintance! I have seen children do just the same thing, so it makes the kinship appear even closer. He shows the most exquisite devotion to his master, caresses him with his pretty baby hands, murmurs *ouf* in the tenderest of human tones, and sits on his shoulder or on his knee as he writes, looking up with a strange wistfulness in his eyes, as if he would like to express himself in something better than a monosyllable.

This is a curious life. Mr Low sits at one end of the verandah at his business table with Eblis looking like his familiar spirit, beside him. I sit at a table at the other end, and during the long working hours we never exchange one word. Mahmoud sometimes executes wonderful capers, the strange, wild, half-human face of the siamang peers down from the roof with a half-trustful, half-suspicious expression; the retriever lies on the floor with his head on his paws, sleeping with one eye open, always on the watch for a coveted word of recognition from his master, or a yet more coveted opportunity of going out with him; tiffin and dinner are silently served in the verandah recess at long intervals; the sentries at the door are so silently changed that one fancies that the motionless blue turbans and scarlet coats contain always the same men; in the foreground the river flows silently, and the soft airs which alternate are too feeble to stir the overshadowing palm-fronds or rustle the *attap* of the roof. It is hot, silent, tropical. The sound of Mr Low's busy pen alone breaks the stillness during much of the day; so silent is it that the first heavy drops of the daily tropical shower on the roof have a startling effect.

Mr Low is greatly esteemed, and is regarded in the official circles of the Settlements as a model administrator. He has had thirty years experience in the East, mainly among Malays, and has brought not only a thoroughly idiomatic knowledge of the Malay language, but a sympathetic insight into the Malay character to his present post. He understands the Malays and likes them, and has not a vestige of contempt for a dark skin, a prejudice which is apt to create an impassable gulf between the British official and the Asiatics under his sway. I am inclined to think that Mr Low is happier among the Malays and among his apes and other pets than he would be among civilised Europeans!

He is working fourteen hours out of the twenty-four. I think that work is his passion, and a change of work his sole recreation. He devotes himself to the promotion of the interests of the State, and his evident desire is to train the native rajahs to rule the people equitably. He seems to grudge every dollar spent superfluously on the English establishment, and contents himself with this small and old-fashioned bungalow. In this once disaffected region he goes about unarmed, and in the day-time the sentries only carry canes. His manner is as quiet and unpretending as can possibly be, and he speaks to Malays as respectfully as to Europeans, neither lowering his own dignity nor theirs. Apparently they have free access to him during all hours of daylight, and as I sit writing to you or reading, a Malay shadow constantly falls across my paper, and a Malay, with silent, cat-like tread glides up the steps and appears unannounced on the verandah, on which Mr Low at once lays aside whatever he is doing, and quietly gives himself to the business in hand.

Isabella L. Bird, *The Golden Chersonese and the Way Thither*, John Murray, London, 1883, pp. 306–9, 321–3.

37
Government Resthouses in Perak in the 1930s

GEORGE PEET

Once the traveller left the larger towns behind him he relied mainly on the government resthouses for overnight accommodation. In every town which was the headquarters of a district, the government built a resthouse, furnished it, and let it to a tenant, usually Chinese, who undertook to provide food and lodging at approved rates of charge. The system was intended to provide accommodation for officials on tour (or awaiting the allocation of a bungalow as permanent accommodation), but, if rooms were available, the resthouse manager was expected to cater for other travellers.

George Peet was for many years on the staff of the Straits Times *newspaper; in the 1930s he was their correspondent in Kuala Lumpur with a roving commission which took him on journeys through the Malay States. He rose to become the editor of the paper after the war. He here describes the rather precise differentiation of classes in pre-war Perak. As he hints, the 'Raja Resthouse' was not really for Malay Rajas but for senior officials who did not care to mix with lesser lights, whether official or non-official. On this occasion his accommodation had been arranged by the Incorporated Society of Planters, whose annual meeting he had come to report.*

MY resthouse at Taiping, I would have you know, was not the ordinary, common-or-garden one. That you will find in the centre of the town, with the date '1894' inscribed upon it, and it accommodates the *hoi polloi*, the people whom the newspapers describe as the 'general public' when they have to draw a delicate distinction between the great ones of the earth and the mass of mankind.

My resthouse was the aristocratic one, reserved for ... well, I don't quite know whom it is reserved for. It is known as the 'Raja Resthouse', but I could find no rajas in the visitors' book. On the contrary, this dog-eared tome was full of the incomprehensible initials with which the officials of our local Lilliputia designate their various functions.

Still, it certainly is a special resthouse. Not everybody can get into it. The vulgar herd is excluded. Between those who go to the ordinary resthouse and the favoured few who go to the superior one a subtle difference exists. What it is I don't know, but it certainly exists. One hesitates to appear at the Raja resthouse in a vulgar rickshaw. Nothing less than an Armstrong Siddeley, possibly with the Straits Times houseflag on the bonnet, seems to be expected.

Staggeringly polite Malay servants take the place of the saturnine Hylam [Hailam Chinese], and the poor journalist wonders how on earth he can keep pace with rajas in the matter of tips. The food is awful, I must admit, and the fish served to me had certainly been snared in the mud of the neighbouring lake.... The Hylam beats the Malay at cooking and the Malay beats the Hylam at good manners, so you must take your choice. But I choose dignity and exclusiveness every time. A taste of aristocracy is too good to be missed, even if one receives it only in a resthouse....

The Raja resthouses are an institution peculiar to Perak. We certainly don't have them in Selangor, nor, I think, are they found in any other Federated State nor in Johore. There is a separate chiefs' resthouse at Pekan, but that is much inferior to the ordinary resthouse, which is a wondrous place with glass windows, a mosquito-proof room and tiled bathrooms. But Pekan after all cannot be compared with other Malayan towns. It is a sort of local Sandringham, the only one in this country where a white man loses prestige if he rides in a rickshaw (as a very conventional young M.C.S. [Malayan Civil Service] officer indicated when he spotted me committing this faux pas during the coronation festivities).

There is also a Raja resthouse at Kuala Kangsar, and I remember being much impressed by the nonchalance with which a friend of mine, one of the lesser lights among the staff of the Chenderoh dam, took me in there one evening and ordered a stengah [whisky and soda]. These customs of other States are not easy to follow, but no doubt Perak folk have no difficulty in interpeting the social nuance represented by the existence of two Government resthouses in the same town.

But if you are an uninformed visitor to Taiping without the proper sponsorship, why risk it? Nothing could be more undignified than to be ejected from the Raja resthouse on the ground of insufficient personal importance. Besides, one can enjoy a visit to Taiping very well without staying there. Admittedly the peacefulness of the situation outside the town, the outlook on the public park and the view of Taiping's grand mountain background are addtional pleasures of no mean order; but they are not essential. Taiping is certainly a town to see if you get the chance, even if you stay at the ordinary resthouse in town.

G. L. Peet, *A Journal in the Federal Capital*, Malaysian Branch of the Royal Asiatic Society, Singapore, 1983, pp. 145–7.

38
Where Women Vote—Negri Sembilan in 1912

RICHARD WINSTEDT

Negri Sembilan means 'Nine States'; it is a confederation of smaller units, each with its traditional political hierarchy of district, clan, and sub-clan headmen, chosen by election from lineages in which descent and therefore membership is determined by the female line; that is, a man is a member of his mother's group, and his children are members of his wife's group, although it is the men who hold tribal office.

Settlers from the great Minangkabau kingdom of central Sumatra brought this distinctive tradition, based on many rules and sayings, with them when they settled in what became Negri Sembilan in the seventeenth and eighteenth centuries. Sir Richard Winstedt, one of the most learned authorities on Malay history and culture, was District Officer, Kuala Pilah, at the time of the 1914–18 war.

ON my return from my first furlough I was transferred from Perak to Negri Sembilan, that delightful little State of lost causes and incredible beliefs, breathing from clustered hamlets and sequestered rice-fields the last absurdities of the matriarchal system. Here in the placid valleys the marriage of your son to your brother's daughter is unobjectionable but his marriage to your sister's daughter is incest; and here a man comes too near a girl that comes to have her snatch his hat or his handkerchief, for she can produce it before the headman as factitious evidence of wooing and sue him for breach of promise. Muslim laws of inheritance, Muslim laws of evidence, Muslim rules of purdah: with such man-made inventions the women of the Nine States will have no truck, and as they stick closer to their land than to a husband, they have introduced a form of marriage with foreign Malays allowing a wife divorce if such a negligible excrescence on her tribe is absent six months on land or a year overseas. No shy reticence marks these suffragettes....

In other Malay States the men visit government offices, and if they have to bring their women-folk, bring them veiled and demure. There is order and decorum and no jostling. In Kuala Pilah my office was always thronged with eager, loud and persistent ladies, who of all the Peninsular Malays are the only people to have invented a word (*terkilan*) for *discontented.* That no patriarchal state has the word may be due to the creese [kris] having formed as ordinary a part of a man's dress as the duelling sword of our ancestors' costume. The women never having worn arms are less reserved and on the very first day of my arrival at Kuala Pilah, *terkilan* was reiterated in my ears like a refrain from an anthem. In Negri Sembilan alone if a man tries to present a case, his wife or sister pushes him aside: 'the land is mine, Tuan, and he knows nothing about it.' One day when I excused myself to a headman for some irritable gesture, he replied: 'That is nothing. A former District Officer got so exasperated one morning that he leapt on to the office table and hurled an inkpot at a crowd of chattering women. He was quite justified; even poultry are kings in their own place. A woman should not pound rice in a pot or boil it in mortar; the Tuan's office is no place for wrangling, and an inkpot is a proper office weapon.'

There are many tribes and sub-tribes—with names like the Banyan-tree Hamlet Sub-Tribe of the Flat Plain Tribe, and every one of them has a headman, whose business it is in tribal language to push open the door of the district office as the prow of a ship cleaves the waves. Most of these vessels of tribal wisdom are old and, as another lay expresses it, the valleys often grow too steep for their feet, so that elections occur almost monthly. Women are the electors and in my time there was no more damned nonsense about merit in the candidates than there used to be about the merit of those designated for the Garter. But the ladies had a freakish sense of humour and could on occasion be joked out of their nepotism into lightly voting against a hurt and indignant relative. At one election I turned the scales completely against their favourite by remarking that the thigh is nearer than the knee, a proverb which revealed the inscrutable white man to be human after all, to be somehow aware that they were backing a relative.

'His skin is the colour of a boiled prawn,' one gap-toothed absent-minded elector observed in a loud whisper, 'and he looks too young to know anything about custom, but he has wit, poor fellow. I have known Dollah all my life and know he is a fool: this young white man has known him a week and has seen through him. Dollah will never be able to cleave his office door. I shan't vote for the nincompoop.'

The supreme joke of this elective system (as of the League of Nations) was that no conclusion can be reached, unless all the voters were unanimous or at one 'like a round jet of water spurting out of a bamboo conduit'. To the student of custom this was amusing, but the harassed District Officer, surrounded by vociferous women, felt like that cucumber among thorns, until a few years ago it was enacted that after six months of disagreement a disputed election would be settled by the Ruler in Council.

Richard Winstedt, *Start from Alif, Count from One, An Autobiographical Memoire*, Oxford University Press, Kuala Lumpur, 1969, pp. 142–4.

39
Kelantan in 1888

CARLO FERDINANDO BOZZOLO

Bozzolo had left his native Italy to work as a dredgemaster under De Lesseps in cutting the Suez Canal. He had been in Perak since 1880 and in 1888 was District Officer, Upper Perak, whence he made an expedition overland to the

Kelantan River and then travelled downriver to Kota Bharu to buy elephants. It was said that he 'writes in an English of his own invention'.

The opening paragraph of this passage refers to the typhoon of 1880 and other disasters which had devastated and depopulated Kelantan. The Ruler of Kelantan at this time was Sultan Ahmad; to him and to his Siamese overlords the arrival of a British emissary was unwelcome. Bozzolo's ccount reflects the hostility on both sides.

I should like to give here a description of the present state of Kelantan. It is calculated that this country at present contains only one-third of its population of six years ago; its decline first began by the burning down of Kota Bahru, then a typhoon which destroyed a number of villages and caused tremendous damage all over the country, then came the cattle disease which swept the country of all its cattle: the last named may be said to be a sort of death-blow to the whole of Kelantan, owing to which all the lands were left uncultivated. The population living in Kota Bahru having many resources of living suffered not so much, those residing at the banks of rivers managed with difficulty to support themselves by getting up rice supply by boats, but the majority of those living inland (called '*orang darat*') were left to the mercy of Providence; the poorer class, after a hard struggle for life, then began emigrating, leaving their families behind, others given to thieving etc. Cattle which were brought and introduced from other States were stolen and killed for food. Those who had padi lands and others who had means of cultivating padi had to give up planting as their crops were stolen before the harvest time came. The Sultan taking advantage of these hard times commenced to lend out money but at a very high rate of interest and from this source the number of his slaves increased: those who had means also imitated the Sultan's way in lending out money but only for short periods, with the result that a number of families, the majority being women and children, submitted to slavery for small sums. The rice which was brought into Kelantan was then the article upon which traders of Kota Bahru made their fortune. Gold ornaments also fell in the market to $15 per tael. Condemned rice, moistened to increase the bulk, also sold in the market from which the poorer class including slaves had to live upon.... For some time the upper class derived a great profit from this source by employing these poor destitute people, some at weaving, others at work in the padi fields, which thing did not last long.

In the year 1885 or thereabouts, as is generally the case, an epidemic in the form of cholera broke out, effecting great mortality principally amongst the poor starving population ... the death-rate becoming so great that the Sultan had several parties of men formed specially told to do the duty of pulling and throwing the dead bodies in the river—during all this while, and strange to relate, the Raja Muda kept up great festivities night and day, *manoras, mayongs* and *wayangs* from different countries were all kept on playing while on the other hand the natives were dying on the roadside....

Any one paying a visit to Kelantan and confining himself only to Kwala Bahru would not discover the real state of things existing at present all over the country. Kota Bahru, the capital of Kelantan, is by itself a very deceitful place, as there will be seen daily the bustle and confusion caused by the number of people going in and out, transacting business, tending to show the prosperity of the country, but directly going a few miles inland may be seen villages after villages, with beautiful houses in it here and there and gardens all abandoned. As an instance of this I will here state that when I was coming down the river, I was surprised to find no one bathing in it during the day, neither any children playing about on the banks, nor a single boat tied up anywhere, but occasionally a few women may be seen here and there coming to fetch water, which indicates that the chain of villages seen on both banks are nearly all emptied....

The weaving industry is entirely in the hands of the women of Kota Bahru, and it is somewhat interesting to notice the preparation of the dye and the process of dyeing which the silk thread has to undergo before it is ready for the loom; although it seems simple and plain, yet I consider it a theory worthwhile learning as to how the different colours are really obtained; to obtain this dye small gardens are being cultivated here and there containing plants from which it is abstracted. The male sex, as a rule, only take up one of the four following professions, viz., fisherman, sailor, carpenter and goldsmith, of which the last two are very numerous, and very able ones too.

Kota Bahru is the centre of all the import and export trade and might as well be termed a town. If properly laid out with large streets, it would be five times the size of Thaipeng [Taiping]; hundreds of sailing vessels of different descriptions are anchored in the river, bamboo houses on rafts and river boats are also very numerous, so much so that at my arrival I was taken aback to see such an important place.

At break of day the ferry boats stationed at various points began bringing from the opposite shore an average of 30 women each trip, keeping going backwards and forwards all day, as well as a number of small *sampans* coming from every direction containing women with their goods for the market. I passed these market places on two occasions during the busiest hours of the day, and I confess that I was very much tempted to stop and buy fruits and flowers which were so well arranged, in particular the latter most neatly made up in small bouquets, and in different shapes which would entice any one to buy. Amongst the variety of fruits I noticed for sale were the water-melon, oranges, limes and mangoes, the last mentioned I found has a superior delicious flavour which I have never before tasted out in the Straits. Every article of food (Malay) that could be imagined may be found every day in the market. A covered market is reserved for the sale of cloth, and from this passing to another which is also covered are brass and iron-ware including crockery and other nic-nacs. Although Kota Bahru with its dirty narrow streets and a number of beggars, gives rather the appearance of a small Arab town, and the only difference is, that no dogs are seen straying about. In the middle of the day a movement begins by those who have completed the sale of their goods going into the river to wash themselves, and what an awful sight it was to behold from my raft the number of men and women mixed up washing themselves in shallow water almost stagnant, with the quantity of dirt floating about them. At about 3.20 p.m. the market begins to close, and almost at the same time the fishing boats commence to arrive landing near the sandbank, in a short space of time groups of hundreds of women are seen surrounding these fishing boats which continue almost till dark. However, at 4.30 p.m. the majority of them begin to start for their homes carrying a small basket containing an assortment of things for their supper, and so before 5.30 the market is altogether abandoned....

The only brick house in Kota Bahru is owned by a well-to-do Chinese, called the Interpreter, all other buildings are of planks or bamboo partition, the Sultan's palace is built all of wood, and it is only the help of Providence that had kept the fire off—its streets are very narrow with heaps of filth here and there, and it is almost a wonder how human beings can live in such a state. The water used by the majority is taken from the river near the bank and it can be imagined how impure it is when thousands of the people make their latrines in the river.

During my stay in Kota Bahru I have seen very little of loading and unloading from sailing vessels, and judging from this I should say the trade in Kelantan is very insignificant at present; the steamer which was then running every fortnight was ample enough to carry all the products of export as well as import.

Carlo Ferdinando Bozzolo, report of a visit to Kelantan in 1888, enclosed with Straits Settlements despatch of 31 January 273 to the Colonial Office (unpublished).

40
Kelantan in 1899

WALTER SKEAT

Kelantan and Trengganu (Passage 41) were formally dependencies of Siam until 1909 and its government might have objected to a visit by Skeat who was a British government official from Malaya (see Passage 5), but at the outset of his expedition he had been to Bangkok and satisfied the Siamese authorities that the purposes of his expedition were scientific, not political. Although the other members of the expedition were zoologists and botanists, Skeat himself was an anthropologist, who later wrote two authoritative books on Malay and aborigine cultures. He, and the other joint leader of the expedition, Dr Laidlaw, were in Kota Bharu, where they made 'measurements' of the physique of local Malays for the purposes of physical anthropology. Luang Phrom was one of the Siamese officials stationed in Kelantan at this time.

8 October.

I visited the copper-smith, to see what I could make of his methods. In the afternoon, Laidlaw accompanied me to the school, where we found seventeen boys present out of twenty-four on the register. There had been a falling off in the attendance, owing to the smallpox, but a proper register—and that in ink—was being kept, and the room was neat and clean. Several boys were set to read for our benefit, three standards in Malay being covered. If none of them were brilliant, all were fairly good, and we left with the feeling that the master had done his best in difficult circumstances. . . .

10 October. The morning was spent by Laidlaw and myself in taking measurements of Kelantan Malays. In the afternoon the Schoolmaster paid us a visit, and told us that a man had been flogged publicly the previous day for an alleged attempt 'to poison two women'. . . . The Malay magistrate claimed to be able to decide all the cases 'by the light of nature'. The schoolmaster said that when he was appointed to Kota Bharu (I believe about four years ago), he got a copy of Clifford's *Translation of the Penal Code* for the benefit of the Kelantan Magistrate's court. But this gift was returned to him the following day. Life in Kelantan (he added) even a short while previously was much less safe. In those days, 'every peasant carried his *keris*, and the roads were quite unsafe after dark'.

After lunch we crossed the Kelantan River in one of the royal boats, a beautiful sky-blue in colour, to see the traditional execution grounds at Pasir Pekan, on the north bank, where for some years all judicial death-sentences had been carried out. Two forms of execution were used, both apparently older than any existing records. In the first (*pujut*), a rope was passed round the condemned man's neck, and taken in a clove-hitch at the back. In the second (*kilas kasau jantan*), a ring of rope or rattan was twisted into a figure-of-eight and pressed against the throat: two stout sticks, about three feet in length, were then passed through the two loops of the noose, with their ends crossing at the back of the victim's neck. Two or more men pulled on each of the rope ends in the case of the first, until the man was strangled. In the *kilas kasau jantan* two men, one on each lever, forced the ends apart until the pressure on the loop broke the cervical vertebrae. It is said that when Sir Hugh Clifford witnessed one of these executions he was sick with horror.

The use of both methods was defended by the Malays at Pasir Pekan on the grounds that they took no longer than would be required to squeeze the life out of a squirrel, and that they had never failed, unlike our own 'humaner' methods. The body was usually carried off by the victim's relations, but some were left lying on the sand till the bones were picked by the vultures. One 'Che Ali had been executed here, we were told, only two months previously. He had stabbed a Bomor* (*Mindok*) in the traditional manner, by thrusting a spear from below through the raised bamboo flooring of a Malay hut, while his unfortunate victim sat above in the 'trance' phase of one of his performances.

On leaving the Execution grounds we landed on the further banks of the Kelantan river, and entered a small Kampong, where seeing more than usually attractive children, I jokingly asked if they would sell me one. The reply was that if I had come there 'two or three months ago' I could easily have purchased several. On my enquiring what this meant, the people of the house informed me that during the late smallpox epidemic the up-river Malays had in a number of cases set their sick children adrift on rafts, with no more than a plateful of rice, and a small vessel of water. At several places these dying children had been washed ashore by the current. Two children had been cast ashore at Pasir Mas, a third at Pasir Kankong, a fourth at Bukit Pandai. Luang Phrom, with whom I subsequently discussed this horrible practice, said that several other cases had been reported in the week before I arrived at Kota Bharu. He had himself taken the matter up with the Sultan at that time, and the latter had expressed his abhorrence of the practice, but did not know what he could do about it as it was not possible to tell from which villages the rafts had come.

*A *bomor* (or *bomoh*) was a Malay doctor using magical as well as herbal remedies. Probably 'Che Ali suspected him of putting a spell on him.

W. W. Skeat, 'The Cambridge University Expedition to the North-Eastern Malay States, and to Upper Perak, 1899–1900', *Journal of the Malayan Branch of the Royal Asiatic Society*, Vol. XXVI, Pt. 4, 1953, pp. 113–14.

41
Kuala Trengganu in 1899

WALTER SKEAT

Skeat here continues his account of a tour of the East Coast States. Sultan Zainal Abidin III was at this time in the middle of his long reign (1881–1918). The remarkable economic progress of his State owed much to the energy of his predecessor, Baginda Omar (1839–76) (see Passage 13 on Vaughan's visit in 1846).

25 October.

WE reached Kuala Trengganu at 6 a.m. on 24 October. This morning (25 October) we had our first interview with the Sultan. He appeared to be about forty-five years old, with open, not unpleasing features and a slight moustache. He possessed more dignity, and seemed to be more at ease, than the other sultans whom we had seen. He was dressed in a loose, cream-coloured jacket, a sarong and white trousers of the usual pattern. On his head he wore a black *songkok* [fez], in place of the elaborately folded head-cloth. On entering, and again at the end of the interview, he shook hands with us in the European fashion. We conversed seated on bentwood chairs on the dais of the *balai* [hall of audience]. While we were there, as at Kelantan, coffee and cheroots were brought to us. The Sultan began by asking our names, and making certain personal enquiries. Then he discussed several general topics, a little remotely perhaps, but he certainly seemed interested in our comments on the wild animals of the districts we had visited. At this point he sent an attendant out; after a little time the man came back with a copy of Wood's *Natural History*, in which we were able to pick out illustrations of some of the beasts we were discussing. Throughout, though quiet, he showed an original mind and a shrewd sense of humour.

This was the Sultan who, when asked why he did not provide free rice in his gaol, as the other sultans did, replied that if that were done in Kuala Trengganu, the entire population would clamour for admission. It was also said that he was once asked why, with all the wealth at his disposal, he did not pay off his debts; to which he answered that so long as he remained indebted, his affairs and his health would be matters of some consequence to his creditors, and he could be sure that they would pray sincerely for his continued survival. His policy may well have been sound on this point, for he reigned thirty-seven years, and lived to see Trengganu pass into the British sphere of influence, and even the appointment of the first British adviser to his State, J. L. Humphreys, in 1915.

Shortly after our return to our quarters a message arrived to the effect that the Sultan had deputed one Che' Ta'ib, of Kampong Losong, to be our guide during our stay. In the course of the next few days he took us round all the many kampongs of which the capital was composed, and we saw all manner of crafts being followed from boat-making to embroidery. Kuala Trengganu appeared to be a hive of industry, and we were truly astonished at the range of the activities, and in parts at the high quality of the work. The only thing that was a little disappointing was that in some cases the material used seemed to be much inferior to the skill and care that was being lavished on it.

The shops, we noticed, were well supplied with bread, light beers, soda, cheroots and similar European wares, as well as with an extensive assortment of Malay goods and, above all, Chinese and Indian articles. The streets, except around the *Istana* [palace], were ill-kept and destitute of drains. Yet even here the local talent for craftsmanship showed itself clearly in the well-built bridges across the creeks. We also saw a few wooden lamp-posts, which must have been unique among the east coast states at the time of our visit. Passing through the kampongs we found that the gable ends of the palm-thatched houses were decorated with many different designs. Some had a sort of half-open gable end, called *alang-buang*, or *undang-undang*. The windows, rarely plain openings, were for the most part barred: some were diamond-shaped, others had alternate pieces omitted in the bamboo plaiting and these made up the design.

28 October. In the morning we visited the famous local shrine (*keramat*) of To' Sheikh Ibrahim, of whom it was told that when the Maharaja of Johore bombarded Trengganu he walked up and down the beach (everyone else having fled to the hills), and by the power of his magic gave the attackers the impression that a numerous army was prepared to defend the town, with the result that they retired without effecting any landing.* This is, of course, somewhat at variance with recorded history, but I give the local 'wonder' story, as it was told to us. No votive flags were to be seen at this shrine, but incense (*haru*) was apparently burnt there, and a goat sacrificed by the old care-taker every Friday. The grave had five posts (*batu nesan*) at each end, making ten in all, instead of the usual single post; the superfluous ones had been added out of the funds provided by the saint's many devotees. To them also, presumably, was due the fact that it was protected by a triple mosquito curtain, and an atap roof-shelter was built over it.

According to Che' Ta'ib the chief difference between the marriage customs of the east and west coasts of the peninsula was that the Trengganu people held their wedding processions at night, instead of at 5 p.m. as on the west coast; and that for this reason their weddings usually took place during the bright half of the moon.

29 October. Che' Ta'ib told me that if a Chinese was detected in an intrigue with a Malay woman, he was arrested and given the choice between becoming a Muslim (*masok Melayu*) and marrying the girl, or going to gaol. If a Trengganu woman was caught in fornication (*buat jahat*) she received forty-four stripes, which were administered in public. Hence public women were forced to go about their trade circumspectly, unlike those of Kota Bharu (Kelantan) where they were bolder than anywhere else of the east coast, or perhaps in Malaya.

*It was the British navy which had bombarded Kuala Trengganu in 1862, not the Maharaja of Johore.

If a bullock damaged a rice-crop, the bullock could be detained by the proprietor of the field until its owner paid the fine, which was assessed according to the damage done. If the offending beast were a buffalo, the fine was doubled; if it were a goat, the fine could not exceed twenty-five cents. These were also the fees that had to be paid if the animal were taken to the pound-keeper (*petanda*). But the owner of the damaged field, or other property, could always accept a smaller sum, if he liked, if the beast was not impounded.

31 October. I paid a final visit to the *keris*-maker, who was preparing a blade for me. I saw it taken out of the sulphur bath, and cleaned with limes; and then was able to take it to have a sheath made for it of the ordinary kind used in Trengganu.

On our way home this morning we noticed men bringing sand up from the shore and depositing it in cloth bundles at the side of the street. The Sultan had ordered the fishermen at the Kuala to do this, because he wished to improve the road. Here and there, in other places, the roads showed signs of having been widened already. The wooden creek-bridges were (as has been stated) good, and there were even two or three small brick-and-stone bridges, which was more than we had seen elsewhere on the east coast.

W. W. Skeat, 'The Cambridge Expedition to the North-Eastern Malay States, and to Upper Perak, 1899–1900', *Journal of the Malayan Branch of the Royal Asiatic Society*, Vol. XXVI, Pt. 4, 1953, pp. 121–2.

House and Home

42
A Villager's Home in the Malacca Countryside

JAMES LOGAN

James Richard Logan is remembered as the founder and editor of the Journal of the Indian Archipelago *(1847–59) to which he made numerous contributions. He was the pioneer in the study and recording of local history and custom. This passage comes from his account of a five-day visit to the interior of Malacca Settlement, then hardly known to the Europeans of the Straits Settlements towns.*

The conventional structure of a Malay house is here described. It stood above the ground 'on stilts', and was approached by a short ladder or flight of steps. The outer room, in the form of a veranda, was the place for receiving guests; the larger interior room was for family living and sleeping; the kitchen was at the back.

ABDULRAHMAN's house lay in one of the plantations on the right side of the road.... The large room into which we ascend from the verandah, is only used as a reception room on feasts and other great occasions, and ordinarily forms a convenient store-room for the less valuable household stuff, such as baskets of different kinds, mats, etc. Around a wooden post in the middle are hung an abundance of spears, swords, and other weapons of several sorts, for the Malayan armoury displays a motley and curious assortment of weapons. A number of baskets of paddy, which had been newly brought in from the field and were not yet cleaned for

Malacca village house, from Phillip Gibbs, *Building a Malay House*, Oxford University Press, Singapore, 1987.

the granary, were placed on the floor. The smaller room was my host's bed chamber, the only place in the whole kampong sacred to privacy. At one end was a curtained bed, and on the other were stuck or suspended some fire-arms and a great variety of krises, swords and knives. Some of the krises were sinuous in shape and damasked or striated,—slight rough ridges rising from the surface of the blade and giving it the appearance of a number of thin plates having been welded together and their edges left projecting. Amongst these weapons were the kris panjang, k. sampana, k. sapukul, chinangkas, klewang, pidang menangkabau, golah Rambau etc.

Abdulrahman pointed out a kris of celebrity, which had gleamed in many a foughten field in days of old and, as he said, 'drunk much blood'. This he regarded with a look of veneration, and prized as his most valuable possession. It is not simply from virtu that a Malay collects and cherishes so many different species of weapons. When the field was the grand source of distinction, arms which had served their owner well in his hour of need were held in high esteem by him and his descendants, while those worn by champions distinguished for their prowess, acquired the repute of being indued with the supernatural quality of giving invincibility to their possessor. As the lapse of time removed from around the memory of a warrior all the more vulgar attributes of humanity, raised him into an impersonation of heroism, and connected his deeds with the invisible powers who had favoured him, his charmed *kris* (*kris bertuah*) became environed by a spiritual halo in the imagination of the Malays....

The owner of all these possessions, and of the paddy fields in front, welcomed us at his gate, and struck me by his abrupt and homely manner, so different from that of the Malays in the town. A good deal of this I found afterwards to be peculiar. His character is plain, direct and in a remarkable degree energetic. He is, compared with many, rude and little tinctured with the pedantry of Islamism, but endowed with strong practical sense, getting at once as deeply into the heart of a subject as his mental range enables him to go, without beating about the bush. At first his manner was embarrassed and apparently dry, and his efforts to break through the restraint under which he laboured were abrupt and highly grotesque. When we ascended into the verandah he blurted out his welcome again, jerked his head about, bent his body forward, and shifted his position every second. He was most delighted, he said, highly honoured, but oppressed with shame. His house was such a miserable hut, and he was such a poor, ignorant, vile person, mere dung in fact! 'Saya orang meskin, tuan,—orang bodo,—tai,' and so he continued vilifying himself, and accompanying each new expression of humility by a sudden and antic alteration of his attitude and position. An ample repast of boiled rice, fish, etc. was soon spread on the mats, and I now learned from Mahomed that our host had left his house in Malacca the preceding evening, and walked 18 miles during the night to have breakfast ready for us at an early hour. The Malay coolies who had been employed in carrying my baggage sat down with my host and Mahomed. A separate array of dishes was provided for me at a little distance upon another mat, and I was invited to occupy the only chair that the house afforded. As the chair was ricketty, and table there was none, I preferred following the custom of my neighbours.

James Richard Logan, 'Five Days in Naning', *Journal of the Indian Archipelago,* Vol. 3, 1849, pp. 27–9.

43
A Village in Negri Sembilan in 1875

ARCHIBALD ANSON

From November 1875 to about March 1876 there was some rather ineffectual Malay resistance to British intervention in Perak, Selangor, and Negri Sembilan. The Governor, Sir William Jervois, was preoccupied with the situation in Perak, where the British Resident had been killed, and so it fell to Anson, as his deputy—and a regular soldier by profession—to take command of the small force which advanced over the hills from Sungei Ujong to the Terachi district in the interior of Negri Sembilan. (On Anson see Passage 30 and on Negri Sembilan custom Passage 38).

THE inhabitants having deserted their houses, and fled into the jungle, I sent messengers to try to induce them to return; but only the women, to the number of about two hundred, came in. They were all brought up to me, and with children, one by one kissed my hand. Those who had infants in arms put them forward to do so also. I then had the women collected together, and addressed them through my interpreter. I asked them why they ran away. They said they were very frightened. I said, 'Do we look such dreadful people that you should be afraid of us?' Again they said they had been very much frightened. I then told them that English people did not fight against women and children, nor against any people who were peacefully disposed. That we wanted rice and other things, and were quite willing to have paid for them; but that as every one had gone away we had to help ourselves, and were unable to pay for them. I then warned them that all their children, that should be born in future, would have long legs, because they ran away when we came. At this there was a general giggle, and we chatted and became good friends. I cautioned them to keep in their houses, away from the Chinese coolies, while we remained.

I occupied one of the best huts (such as it was) in the village, and made the acquaintance of a very old man, who had in his youth come from Menankerbow [Minangkabau], on the west coast of Sumatra—supposed to have been the original country of the Malays. I invited him to come up the usual rough ladder, of four or five wide apart rungs, into the main room, and chatted with him; and, desiring to entertain him, offered him the only thing I had available at the moment—some mixed Reading biscuits; but alas! he had no teeth, and could not eat them. He became very friendly, and told me where I could find some brass guns, which I sent for, and had brought to me. In the meantime Colonel Clay had sent his men to search for guns, and had had no success; so I pointed out to him the superiority of diplomacy, and a box of Reading biscuits, over military force.

The owner of the hut I occupied at Terachi complained to me that his cocoanut monkey had been stolen. Now, a cocoanut monkey is a valuable property, as it is trained to ascend a cocoanut tree, with a long string attached round its neck, to throw down a cocoanut. It seizes a nut, and if it should not be the one required, a jerk of the string by the person holding the end of it induces the monkey to let go the nut and seize another, and so on until it seizes the required one. It then clasps the top of the nut, which with its outer husk is of great size, with its two little hands, and the bottom with its two little feet, and proceeds to twist it round until it falls to the ground. As the stalk is of considerable thickness, and very like rope, this is hard work for the short, curly-tailed little chap. I asked the Malay if he could describe the person who had taken his monkey, and he gave me a very accurate description of one of the commanding officers. I then wrote to that officer, stating that I had been informed that one of his regiment had stolen a monkey, and suggested that he should find out the man and have him flogged. The monkey came back to its owner very quickly.

A. E. H. Anson, *About Others and Myself 1745–1920*, John Murray, London, 1920, pp. 339–40.

44
Moving House in Kuala Langat

EMILY INNES

We have seen (Passage 6) that on first arrival Emily Innes found that 'the house was worse than I had expected'. Its main drawback was its situation on low, swampy ground alongside the river bank at Bandar Langat. However, it was agreed that a new bungalow should be built on Jugra hill, where there was air and a view. But when it was half ready for occupation James Innes was away at Klang deputizing for the Resident, during the latter's absence on leave. Emily would not wait for James's return.

ON recovering a little, I became feverishly anxious to go to the hill for change of air. Two of the rooms in the new house were now finished, and there appeared to be no immediate chance of the rest being built, so I would wait no longer. The natives tried to dissuade me from going, saying that it was not safe; that the hill was full of tigers, and that to live in a house in the middle of the jungle without a fence round it was like asking the tigers to walk in. To all this I replied: 'If I stay here I shall die of the swamp; I would rather go to the hill and risk the tigers'; and I wrote to Mr Innes to this effect. He had wished me to remain at the Bandar until the new house was finished, by which time he hoped the Resident would have returned, and he himself would be free to superintend our flitting; but finding how much I was set on getting out of the swamp at once, he gave way, and sent me a big cargo-boat to carry the furniture.

The boat arrived late one evening—too late to do anything that night; but the mere sight of it acted as a cure on me, and next morning I was up at five, packing and superintending servants, boatmen, and police, who all helped willingly, but without much judgment. I never worked so hard in my life as I did that day. I was fifteen hours without sitting down, and almost without food, for I dared not let the men do anything without showing them how. They were terrible fellows for making a mistake wherever possible. Having made them take the mosquito-house, beds etc., to pieces properly, I had turned my back for a moment to attend to something else, and only turned round again just in time to save my unfortunate piano from being taken to pieces, note by note, and string by string. The men were, I suppose, acting on the principle that what is sauce for the goose is sauce for the gander. In another half-hour, had I not interfered, my piano would have been in a hundred pieces, and carried thus carefully to the hill in red pocket-handkerchiefs knotted at the four corners.

We were obliged to do everything in one day, as the cargo-boat was wanted back in Klang immediately. At last the final trip to the hill was made, and the piano and I were landed on the rock. I sent the cargo-boat back at once to Klang, and then set myself to the task of getting the piano up the hill. The Malays found it so heavy that they were for leaving it on the rock all night, exposed to the chance of tropical storms; but I insisted on its being carried up at once. I felt really sorry for the poor fellows who had to carry it. It was only a cottage piano, but being made expressly for a tropical climate it had an unusual quantity of brass in it, which made it very heavy. It took about forty Malays to carry it up the hill; they slung it on two long bamboos, ten men grasping the end of each pole in front, and ten more behind. Their mode of going up the hill was to take a rush of two or three steps and then a rest, all the way up; uttering, as is their wont when carrying trees or any heavy weights, the most awful shrieks and groans. Several times I ran out of the bungalow, where I was awaiting them, in alarm at these cries, fearing the piano had fallen on some of them, and at least broken some dozens of their legs; but eventually both the piano and its bearers reached the house in safety.

There was still much to be done before rest could be thought of that day; the Government safe had to be secured by chains passed through holes sawn in the floor; a bed had to be put together for me and completely prepared before the rising of the mosquitoes at sunset (otherwise it was in vain to hope that the curtain would be free of them), and various other things had to be unpacked and made available for cooking and serving up something to eat. At last all was done, and at about eight o'clock I dismissed the police and boatmen, with the exception of two policemen on guard, and had my large, new empty house all to myself.

From that day my health improved in a surprising manner.

The situation of the new house was lovely, and extremely healthy. It was built on a spur of the hill of Jugra, the river winding about below, disappearing and reappearing through the trees, and the hill itself sloping away above the house at the back. We had, months before, begun to make a garden, and one of my constant employments was helping the gardener—for we were now allowed a gardener by Government—by my criticisms to make a really flat lawn for lawn-tennis. I used to stoop down with my eyes on a level with the plateau which was being formed on the side of the hill, and suggest that there was too much earth here and too little there, until at last I got it to my satisfaction. On Mr Innes's next visit he was delighted to find how much the house and garden had progressed since my taking up my abode at the hill. Not long after that, the Resident returned to Klang, and consequently Mr Innes was able to resume his own post at Langat.

Emily Innes, *The Chersonese with the Gilding Off*, 2 vols., Richard Bentley and Sons, London, 1885, Vol. 1, pp. 232–7.

45
A Naturalist at Jeram in 1878

WILLIAM HORNADAY

Before his visit to Kuala Lumpur (Passage 22) Hornaday went up the coast from Klang to the village of Jeram, to hunt for crocodiles (on which see Passage 87 below). Hornaday had a great gift for getting on to friendly terms with the local people with whom he stayed.

THE only house in the hamlet which could afford me shelter was that of Datu Pudeh, the Malay headman of the place, and having been confided to his care by Mr Syers, he took me in, and gave me a corner of his front room, in which I hung up my hammock and *musquitero* without further ceremony. When the tide was in, the house stood almost at the water's edge, rather low upon its posts, with slatted floor, and roof of thatch which had in it several holes large enough to have thrown a dog through. I suppose that, like the man of Arkansaw, when it rained they couldn't fix the roof, and when it did not rain they didn't need to. We had no sooner moved in with our belongings than it began to blow and rain very hard. The bamboo curtains outside were let down over the windows, and the place made as snug as possible, but the wretched old roof leaked like a shower-bath.

A mile above Jerom, a muddy little creek, called Sungei Bulu, runs into the sea between two wide banks of soft mud which are submerged at high tide, and left four feet out of water when the tide is out. A little way up from the mouth is a village of Chinese fishermen who are engaged in catching prawns and making them up into a stinking paste called *blachang*. Every house in the village is tumble-down, rickety and dirty beyond description, and the village smells even worse than it looks. . . .

Before leaving I gave the Datu's wife a very nice figured sarong, which pleased her mightily, and called forth from her most earnest apologies for their inability to entertain me in better style during my stay. She insisted on cooking a hot dinner for me just before we were to start, to which I finally consented, to please both the lady and myself. There was presently forthcoming a very nice and highly palatable meal of fried bananas, preserves of nutmeg and pomegranate, and a dry short-cake to eat with butter and sugar, made by the Datu's mother-in-law. In one sense it was not much, all told, but in another it was a feast, for it was the very best the house could offer.

The mother-in-law and daughter had often peeped through the crack of the door at me, but never had shown themselves until I sent in to the old lady a knife, fork, and spoon as a present, instead of the spoon she had craved as a curiosity; whereupon she forthwith donned her best sarong and jacket, and came into the room where I was, to thank me for her presents and her daughter's. (Nothing makes a man feel meaner than to give a poor present and see it appreciated far beyond its worth.) But her daughter's face I never saw.

William T. Hornaday, *Two Years in the Jungle: The Experiences of a Hunter and Naturalist in India, Ceylon, the Malay Peninsula and Borneo,* Charles Scribner's Sons, New York, 1885, pp. 304, 312–3.

46
Royal Hospitality at Johore Bahru

FLORENCE CADDY

At the end of 1888 the Duke of Sutherland, whose interests included railway construction projects, set off in his yacht to visit Siam, ostensibly to recuperate his health, but in reality to negotiate, if he could, a contract for railway building in Siam (in the event he was unsuccessful). Partly as 'cover' for his real intentions he included in his party Florence Caddy, a writer (and something of a social climber) who had already written books on domestic architecture and on the life of the botanist Linnaeus. These interests appear incidentally in this passage.

On the return voyage the yacht put in again at Singapore, where the Duke and his party accepted an invitation to stay at Johore Bahru as the guests of Sultan Abu Bakar of Johore (in early March 1889). The Sultan was a familiar figure in European society—he had once been entertained by Queen Victoria—and he combined a very European lifestyle with observance of Malay customs, such as abstinence from alcohol. He delighted in showing off his possessions, for example, the expensive dinner service originally made for a Governor-General of India. Florence Caddy's account of the party's stay at Johore Bahru is a unique picture of the Sultan and his entourage.

THE gay little town of Johore Bharu, a Malay town, with an admixture of ten thousand Chinese, centralised by a market-place of architectural pretensions, with a sea-side portico built for the reception of the young Prince of Wales, was festive with flags, and the shipping of small craft in the straits was all gaily dressed with the Johore flag, dark-blue with a red quarter charged with a white crescent and star—but the precise colour does not seem to signify greatly, so that it has the crescent and star, and dark-blue somewhere.

The belt of sea looks like a river, or, rather, like a narrow lake, so blue and smooth. They say it can be rough sometimes. The Sultan's house near the sea appears a comfortable country-seat. Its gardens slope down to the water's edge. It is set in palm-trees and the beautiful ash-leafed tree, poinciana regia, known as flamboyante, or flame of the forest, with scarlet flame-like blossoms, and other trees, some with what we should call autumn-tints in Europe. The leaves do fall, even in the tropics, though imperceptibly, so that but a few trees are bare at a time. Dato Sri Amar d'Raja, the Sultan's private secretary, a highly intelligent young man in European dress, and speaking English fluently, came with another Malay gentleman on board the *Sans Peur* to meet us. The latter gentleman wore the checked silk or cotton skirt, like a duster, round his waist, that is the national sarong.

The Sultan, a stout, pleasant-looking man of middle age, with olive complexion, wearing drab clothes and gold bracelets, received us at the head of the garden stairs of the palace and offered us tea, which was spread on a large round table in the entrance. The view of both shores of the straits from this portico was truly charming.

We were shown our rooms: the Duke and we ladies had each a pleasant suite of rooms apportioned us, with bed-room, dressing-room, ante-room, and drawing-room facing the sea, where we could see the *Sans Peur* behind the palm-trees, and a bath-room below each suite, approached by a winding stair from the dressing-room. Instead of doors there are screens raised eight inches from the ground, fastened at the top, at about six feet from the floor, with a sliding piece of carved wood. This arrangement, only less ornamental, is the custom at Singapore. There are no locks or bolts; it is understood that no one opens a door whose sliding-panel is drawn across. Animals can run in under, but only a very tall man can look over.

The Istana, as it is always called, the Malay word for palace, is European, that is Anglo-Indian in build; in style Renaissance. It was built entirely by Chinese workmen under a European architect. It is internally handsome and well-furnished; the halls and rooms very large and lofty, and the marble staircase broad and fine. The saloon and ball-room on the first floor are hung with rich damask draperies and large portraits of our royal family, and lined with tall Japanese vases, brought home by the Sultan from Japan, with other handsome Chinese and Japanese ornaments; the other furniture, sofas, ottomans, etc. all European. The stair-case and saloon have many tall trumpet glasses, eight feet high, full of tall fronds, or rather boughs, of the delicate phoenix rupicola variety of the date-palm, the glasses twined with climbing fern; this makes a

Maharaja (later Sultan) Abu Bakar of Johor, from the *Illustrated London News*, 4 August 1866.

most elegant and striking decoration, giving an appearance as of a grove of palm-trees with their gracefully waving plumes reared high above our heads, though not nearly reaching the lofty ceiling. The floral decorations all over the house are worthy of the tropics, besides the ferns so bright and green, the various crotons and begonias so rich and dark and velvety, and all so tropically luxuriant as scarcely to be imagined by a Londoner.

As we met each other in a large verandah-like room, common to all of us, between our private apartments, we said, 'We shall enjoy this place thoroughly'; and we all secretly wished our stay might be longer than the two days we had at first almost unwillingly spared.

The Sultan appeared elegantly dressed for dinner, in a monkey-jacket, with the order of the Star of India, and a black velvet fez, with an aigrette of large diamonds in the front; half-a-dozen large gipsy-rings on each hand, almost covering his dark, fat little fingers; the rings all rubies and diamonds on one hand, all emeralds and diamonds on the other. The secretary, Sri Amar d'Raja, was dressed in real English fashion; the other Malay gentlemen wore black coats and trousers, and coloured check sarongs. These Malays were less akin to Europeans in feature than the Siamese, but I cannot see that the Malays are a distinct race from all others; I trace Mongolian features in every line.

I was taken down to dinner by the Sultan, and found him agreeable to talk to. His English is good, though less perfect than that of some of his suite. The very long dining-room is cooled by a line of punkahs, and by open corridors on each side, lined with ferns and other plants. The Sultan bought in London the famous gold dinner-service made for Lord Ellenborough when Governor-General of India, and never sent out. A portion of this was used to decorate the Sultan's table. The large wine-coolers, filled with flowers, are heavy, but the smaller pieces of this service, in Neo-Pompeian style, are very elegant.

After dinner, we had a number of the Sultan's carriages and gharries, and drove through the busy, stall-crowded streets of the town to the Chinese opera, where we sat in a kind of state barn, at some distance, luckily, from the singers, who acted on a raised stage, with a proscenium or frame round it, and simple, fixed scenery. There was a promenade parterre between us and them. The spectators stared at us more than at the other spectacle. More than ever I was struck with eastern costumes as being such a mixture of nakedness and jewels.

The play had a good deal of casting up the legs, and

twirling, strutting, striding, and stalking, as in our barn or fair-theatres, the nature of actors being the same all the world over. The piece was to us like a pantomime, with processions of first, second and third heroes, all equally heroic; alternately with four soldiers going round and round as an army. The voices were mostly in falsetto. The best we could say of the singing was, 'It is a beautiful inarticulate row.' The clashing of cymbals, and thumping of serpent-skin cylinders and drums was a din, and nothing less. In music, the Chinese and Malays are very, very far behind the Siamese, whose music is heavenly compared with this; indeed, it is very pleasing, and often delightful—a real art, and not a discordant screaming and clashing.

We ladies had a carriage, and went home after the opera; the rest waited to see the fire-works, which I heard were fine, and then they went to the Chinese gambling-house, which, it seems, is the chief fun here. Johore is considered an Asiatic Monte Carlo.

Florence Caddy, *To Siam and Malaya in the Duke of Sutherland's Yacht 'Sans Peur'*, Hurst and Blackett, London, 1889, pp. 232–7.

47
'The Virtuous Establishment'

VICTOR PURCELL

Until the Second World War there was a substantial, but diminishing, excess of men over women among the Chinese, reflecting their origin as an immigrant community. Under legislation enacted in 1930, the further influx of male Chinese was severely restricted but no such limits were placed on the arrival of women and children. It was still considered necessary, however, to maintain the safeguards introduced many years before, and enforced by the Chinese Protectorate (a government department), against the risk that women and girls on arrival would be drawn into prostitution or otherwise abused. One of the main pillars of the system were homes (Po Leung Kuk— 'The Virtuous Establishment') in major towns in which Chinese women and children could find refuge, and to which the Protectorate staff could direct them on first arrival, if this was thought necessary.

Purcell here describes the Po Leung Kuk as he knew it (when an officer of the Protectorate) at Penang around 1930.

ON the arrival of each vessel from China the women and girls were taken to an inspection depot. Those whose bona fides were obviously satisfactory were immediately released. Others were released to relatives who came to claim them. If there were any doubt, the Protector might under the provisions of the law require the person into whose charge the girl was going to enter into a bond for a sum of money, usually with two sureties, that she would not be disposed of to others or trained or used for the purposes of prostitution. It was usual to accept shopkeepers of standing and others as sureties. If afterwards the conditions of the bond were found to have been departed from, the bond was forfeit and the girl in question placed in a place of safety—usually the Po Leung Kuk. In cases where there was reason to believe that the girl was the victim of traffic, she was committed to the Po Leung Kuk by the Protector. Here she would remain until suitable provision had been made for her welfare or until she reached the age of eighteen years. Many girls were placed in the Home as a result of raids by Protectorate officers on brothels. (From January to August 1931, out of 266 girls admitted to the Po Leung Kuk in Singapore, 159 were suffering from venereal disease. One of them was 7 years old, another 8, two were 10, and one 11.)

The Po Leung Kuks were established in several centres in the country. That in Penang had been built largely from subscriptions of individual Chinese and was carried on with the support of the Government. It was pleasantly situated in its own grounds and was made as little like an institution as possible. The girls did their own cooking, attended school, and learnt needlework, at which they were usually very adept. They were given outings, visits to cinemas, etc., wherever possible. Sometimes these girls were adopted by respectable Chinese families; more often they remained in the Home until they arrived at marriageable age or went out to work.

The sex ratio being as it was it can be understood that the demand for brides from the Po Leung Kuk was considerable. A man who wanted a wife and could not afford the large expenses involved by finding one in the ordinary way through a 'go-between' would go to the Chinese Protectorate and apply for a girl from the Po Leung Kuk. If he was prima facie a suitable candidate he was questioned as to his name, tribe, occupation, wages, etc. The matter was then passed to the Chinese member of the Po Leung Committee who belonged to the same tribe as the applicant, and the Committee member thereupon inquired into his statements and rendered a report to the Protector. The next step was for the man to attend at the Home. Girls on the 'marriage list' were produced and he made his choice. He had often in view some particular girl whom he had met or heard about in some underground way. The girl selected would either consent to marry him or would refuse him there and then. If she accepted him he had to be medically examined and if the result was satisfactory he had to pay in a sum of 40 dollars for his wife's trousseau. This consisted largely of dresses made in the Home. Then the horoscopes of the two were compared and a lucky day was selected for the wedding. The pair were photographed, the bridegroom entered into a bond to treat his wife well, and they left the Home. Usually these marriages were successful; if they were not, the bride returned to the Home and later went out to work. The usual impediment to successful marriage was some economic factor; for example, squatters and fishermen were unpopular as husbands—the first because they usually lived in the backwoods too far from a town and the second because they were often so poor.

Victor Purcell, *The Chinese in Malaya*, Oxford University Press, London, 1948, pp. 178–9.

48
Moving In at Parit Buntar

KATHERINE SIM

As government officials were liable to be moved about the country, often at short notice and after only a brief period in each place, the colonial government provided them with houses and heavy furniture, but the occupier was expected to satisfy his (and his wife's) requirements in domestic equipment, etc. Usually there were just enough 'government quarters' to house the normal complement of officials at each station. Hence an official and his family, on first arrival at a new station, had to be temporarily accommodated in a hotel or government rest-house until the official whom he was to replace vacated the quarters reserved for the holder of the relevant post. Meanwhile, their colleagues made them welcome, as new members of the local community, with sometimes overpowering hospitality.

Katherine Sim, like many another 'new' wife found it all strange and rather an ordeal. The 'meat safe' was a room walled on all sides with metal mosquito-proof netting.

AFTER my first evening of this I had a splitting headache and felt very depressed. People were extremely kind; but had not the slightest idea how odd this kind of talk sounded to newcomers. I soon learnt it all; although, on principle, we ourselves tried to avoid using much Malay mixed up with English. It sounded worst among the children; some could hardly understand English at all, and called every European woman 'Auntie'. Most parents had become so nauseated with this habit that 'Auntie' was a forbidden word for the amah to use.

A girl fresh from England, hot and tired by a long drive from the port, was greeted with what was practically gibberish to her: 'Will you have your *mandi* now and a *pahit* after?' asked her hostess. The poor girl was puzzled, and no wonder. Why not say bath and a drink. How odd it all was at first! Especially the incredible smells. For the first few weeks your nostrils are so sensitive to the overwhelming pungency you wonder if it is going to be bearable. There are smells of mud, of rotting fish and of ozone; of ripe fruit; of the deep water drains, full of rich black mud and open to the sky along every road. There is the coconut-oil smell of the Tamils, the dry smoky smell of Chinese houses, the acrid bitter-sweet muskiness of the Malays' and the sickliness of the Sikhs' ghee. Then there is the milky-warm stench of cattle and all the minor smells such as that of old rattan, of bat guano and rotting wood, and the peppery dankness of jungle undergrowth. And outsmelling everything is the unspeakable smell of the durian. Those are the bad smells. Mingled with these were heady wafts of pigeon orchid; the sweet cloying scent of the frangipanni flowers; the spicy smoke of joss sticks and smudges; the delicious tang of wood fires and the fragrant, hungry smell of curry cooking.

At first all these odours get up and smite you in an indistinguishable mass, as bewildering as the Anglo-Malay conversation; but in time it was possible to sort them out. And after a few weeks your nose objects only to the really potent ones such as rotting fish entrails on an ebb tide at noon; or a lorry load of durians in the season; or a junk full of copra....

When eventually the difficult young man decided to leave the house empty for us we went along and had a look at it. It gave us rather a shock at first. It was filthy. All the black-painted Government furniture was crammed into one room, the mosquito-proof room or meat safe, as it was called. We laughed at the sight of a particularly hideous long-legged fern-stand which, as the finishing touch, was plonked in the very middle of the room under the fan. Ah Seng simply fell on the house and with a bevy of coolies he cleaned away the first layer of dirt and we started to move in.

The hall and dining-room were attractive, flagged with black and white tiles; and downstairs there was a big room suitable for a studio. The sitting-room and two large bedrooms with bathrooms and dressing-rooms each were on the first floor.

My oak bureau had arrived safely from home, and when it was unpacked two small Tamil coolies tottered up the stairs with it; we could not think why they staggered so agonizingly, until we discovered later that the entire dinner service of fifty-six pieces had been packed in the three drawers....

We had to get in stores, and Ah Seng and Kuki wanted new things for the kitchen. So we went shopping in the town, which consisted of three short streets of semi-Chinese style two storeyed houses, washed azure, viridian green and cream. The shops below were just open rooms or tunnels under the dark arches of the five-foot way; Japanese photographers, Chinese carpenters, a tinsmith and a rattan basket-maker, several Indian silk and cotton stores, goldsmiths and a pawnbroker, and a Japanese hairdresser, Chinese and Indian food shops and petrol stores and Malay coffee shops. There was an open market near the river with food stalls and portable kitchens. Fruit vendors squatted at every corner and spat red betel-nut juice all over the road, so that you had to watch your step. They sold brilliant scarlet, prickly rambutans, golden coconuts, purple mangosteens, yellow pummeloes, slices of pink watermelon and sweet corn on the cob, and above all the great brown stinking durian which is believed by the Malays to be an aphrodisiac.

We ordered all we needed at a sort of general store and four half-naked Chinese carried the cases out, Mat and Ah Seng with a few odds and ends bringing up the rear. We bought curtain material for the incredible sum of 4 ½ d. a yard. Then the carpenter came and made some quite good bookcases and tables from pictures in Heal's catalogues.

In the house we started a battle with coconut spiders, cockroaches and fruit bats. The battle of the 'poochies', as the insects are named collectively, was a never-ending one in Malaya. The wire net of the 'meat-safe' was only enough to keep out the bigger insects and the mosquitoes. The tiny things came through in their clouds at times, in spite of the fragrant smudge sticks burning under the chairs, and became entangled in one's hair. It was not always as bad as it sounds for Malaya is a strange country of variable moods. It is a point in its favour that however unbearable life might seem there occasionally, the gloomy prospect will almost certainly alter within a few hours. Its worst moods work their damnedest with oppressive heat, clouds of infuriating insects, rank smells and discomforts; but as swiftly as the insects descend so the mood lifts. The rain will fall, the air is cooled, the insects disappear, the moon comes out to silver the palm trees and shine on the wet flowers, and once more orchids perfume the air.

There was one particular amphibious bug which, if it fell into water, seemed to be just as happy swimming about as flying in the air or rolling on its back on dry land. So one did not feel safe even in the bath. After six in the evening, when the mosquitoes started, life was a little difficult out of the 'meat safe'. That was one of the reasons why Penang afforded a respite; there appeared to be no mosquitoes there. They were so bad in the Parit Buntar evenings that it was necessary to wear a long *sarong* to protect the feet while one was dressing; and even then one's arms and shoulders were defenceless and were soon covered in large red bites. We had a vast mosquito net; it was like a room within a room, and glowed white and soft when the bedside lamp was lit. There is a comfortable feeling of security—at least from swooping bats and zooming insects—inside a mosquito net, and it is good in the dark to watch a stray fire-fly fluttering against its folds. But the net must be big enough and hang well to the floor; you stifle in the tent sort and the ones that tuck under the mattress are infuriating, as you become entangled in them when you want to get up in the morning.

But a mosquito net only gives you a feeling of security when the light is on. One dark night there was a sinister padding of bare feet apparently in the room. I could not help but think of all the stories I had ever heard of Malays running *amok*, and then Stuart said in a sepulchral voice: 'Don't move, I think there's someone in the room!' I lay sweating with fear expecting to be stabbed in the back at any moment, till Stuart found the switch and the light flooded on. There was nothing there! We searched everywhere, the entire house. But we never found what it was. People said it was a civet-cat in the roof; but it was uncanny all the same.

Fruit bats were another pest but they were something we could cope with. One Sunday morning after breakfast Ah Seng started chasing them with a mop tied to a pole. Stuart joined in and finally eight bats were flying madly about the house; hastily I shut myself in the 'meat safe', and so the landing formed a cul-de-sac. Ah Seng drove a parade of bats upstairs, while Stuart stood on the landing and swiped each one as it arrived. They killed ten between them. During a lull Stuart went to have a cool shower; but Ah Seng with the lust of battle still in him yelled: *Tuan! Tuan!* and Stuart rushed out dripping from the bath armed with the racquet, to see Ah Seng solemnly marching upstairs with his tall pole and one solitary bat flying ahead in the high roof. This went on at intervals throughout the morning.... Later I boasted to a local planter how many we had killed, which amounted to fifteen or so by then, but he said: 'Oh, that's nothing. I got fifty in my bungalow last week.' Ah Seng had thoroughly enjoyed his morning; his hair stood on end and his crinkly smile was triumphant. Usually he was solemn and immaculate in spotless white with smooth black hair, except as now when, as he said, his 'heart beat fast'. Dishevelled he looked most human.

Katherine Sim, *Malayan Landscape*, Michael Joseph, London, 1946, pp. 17–18.

Temples and Mosques

49
Managing a Mosque

DANIEL VAUGHAN

Jonas Daniel Vaughan was a man of many accomplishments. He began his career in the navy, which brought him to the Straits (Passage 13). He then served the Straits Settlements government as a police officer, as Master Attendant (that is, head of the marine department) in Singapore, and then as magistrate and senior local administrator in Singapore until 1861, when he retired and took up a new career as a lawyer, specializing in criminal work. For good measure he was a talented writer, a fine singer, and 'the best amateur actor of his day'. He was drowned at sea, presumed washed overboard from a coastal steamer in 1891. He wrote on both Malay and Chinese customs (see Passages 68, 75, and 86 below). Although this passage brings out the various duties of mosque officials and Vaughan refers to them as a 'priesthood', there was no Islamic clergy in the ordinary sense of a profession apart from its congregation.

THE Musjids [mosques] are either public or private; the former are constructed by voluntary contributions on ground ceded by the Government to the Mahomedan community. A Committee of influential men manage the secular business of the Church. In addition to the voluntary contributions of the faithful, people are allowed to build on the spare ground attached to the mosque by paying a monthly rent; there are also fees levied for the performance of religious rites. The revenue thus derived is expended in supporting the Priests, lighting the Musjids and keeping the buildings in repair.

Private Musjids are built by those who believe that such munificence entitles them to an extra share of Mahomed's protection, and a sure entrance into Paradise.

The highest rank in the priesthood is the Kali [kadi]. He has the power to marry and divorce, in excess of the common duties of a Priest, and may delegate his authority to others as stated above. To enforce the attendance of witnesses and those accused he issues summonses and his orders are generally obeyed with fear and trembling by the poor and ignorant.

Each Musjid has the following personages attached to it—viz., the Imam, Katib, Bilal and Siak.

The Imam leads the congregation in the religious services on Fridays and reads the Koran and other books at festivals. He can perform all the Kali's duties with the latter's permission. The Katib is an assistant Priest with similar powers to the Imam.

The Bilal's duty is to announce the approaching day of prayer on Thursday Eve and early on Friday morning from the Minaret of a mosque. He is selected generally for his loud voice and it is surprising how distinctly he may be heard all over the town calling on all the followers of Mahomed to bestir themselves and spend the day in prayer as all good Mahomedans should. The invitation is readily responded to and from all quarters the young and old may be seen hastening to the mosque to pay their early sacrifice.

The Siak performs all menial duties, attends to the comfort of the devout, sweeps and illuminates the mosque, washes and clothes corpses and assists the priests as required.

An inferior grade in the priesthood is the Lebby or reader; he never conducts any rite but acts as an assistant and may read religious books at feasts and at private residences when required.

The priests and their attendants receive no fixed salaries but are paid from the revenue of the mosque and receive presents at each rite or ceremony from those at whose request it is performed. On the 27th day of the fast month, a collection of rice is made from each householder, and the quantity collected is divided between the Imam and his assistants.

J. D. Vaughan, 'Notes on the Malays of Pinang and Province Wellesley', *Journal of the Indian Archipelago*, New Series, Vol. 2, 1857, pp. 154–5.

50
Celebrating Hari Raya

ZAINAL ABIDIN BIN AHMAD

Zainal Abidin bin Ahmad (generally known as Za'ba) was one of the most influential professional teachers and writers on Malay literature and culture of the first half of this century.

The austere regime of Islam is punctuated by the celebration of a number of festivals, according to the Islamic calendar. Perhaps the most important and widely celebrated of them all, in a social as much as a religious sense, is the feast day (Hari Raya) which marks the end of the month of fasting.

IN Malaya, on both Hari Rayas, all Muslims and Malays in particular, make it a point to visit their parents and religious teachers early in the morning, asking for their prayers and forgiveness. Earlier still that morning the more pious of them had washed themselves ceremonially—an act intended to symbolise the washing away of all past sins. Returning from the early morning visit to parents and religious teachers, they put on their best clothes, and after feasting themselves with the cakes and special dishes prepared for the occasion the previous evening and night and sometimes even during the previous three or four days, the male members of the family hurry to the mosque for the *Hari Raya* service. (The women, however, do not go to mosque for worship on any occasion.)

At the mosque, they all sit in rows on the carpeted floor and chant both individually and in chorus, again and again, special thanksgiving hymns in the Arabic language prescribed for the occasion in praise of God. These are not accompanied by any music. After this, they perform a special *Hari Raya* prayer in congregation, on the conclusion of which the Imam or Prayer Leader ascends the pulpit and delivers his sermons—always two separate ones with a brief interval between them. In these he exhorts the congregation to a more careful attention to their moral and religious duties during the ensuing year, and to do good and shun evil as taught by Islam and other religions.

As soon as the sermons are concluded, every one goes to greet the Imam, shaking (or rather joining) hands with him and congratulating him with the appropriate expression of good wishes. Then follows a general greeting and hand-joining (equivalent to shaking-hands) among the assembled crowd between one person and another, every one offering congratulations to the other, and at the same time asking forgiveness for all past wrongs as well as for any eating or drinking of each other's substance that may have been done during the past without honest and sincere consent. In short, the whole atmosphere for Muslims at that moment, and in fact throughout the day, is full of peace and brotherhood. The very air itself is felt, as it were, to breathe complete harmony and friendliness.

Following the general greeting, the gathering disperses; some go home directly to receive visitors, others go to pay *Hari Raya* visits to friends and relatives. As a rule, Malays keep an open house during the *Hari Raya* for visitors. In prosperous times they usually make special *Hari Raya* cakes and sweetmeats to offer all friends and visitors that drop in.

The children, of course, put on their best and newest dress and go out to their friends' house and elsewhere to enjoy themselves. Naturally, to them the occasion bears no significance except that of enjoyment: they enjoy the food, the cakes, the new clothes and the cheerful company, and also until a few years ago, in imitation of the Chinese, the firing of crackers.

The older people take the occasion more seriously. Wherever friends meet on the streets and in the shops they greet each other enthusiastically with the words *'Selamat Hari Raya!'* meaning 'Blissful Hari Raya to you!' or with the Arabic formula *'Mina'l-'a'idin!'* i.e., 'May you be one of those who enjoy this happy return perfectly!' or 'May you have many happy returns of this day again and again.' The Pakistan

Muslims, on the other hand, say *'Eid mubarak!* which is also Arabic, meaning 'May the *Eid* be a blessed one for you!' But whoever it may be, while they utter these greetings they seize each other's hands (and sometimes embrace each other), or if at a distance they simply raise their hands to the forehead and cry out their greetings. In reply, the same formula is repeated by the friends to whom the good wishes have been directed in the first instance.

Zainal Abidin bin Ahmad, 'Malay Festivals and some Aspects of Malay religious life', *Journal of the Malayan Branch of the Royal Asiatic Society*, Vol. XXII, Pt. 1, 1949, pp. 95–6.

51
Chinese Temples and Processions—Changing Attitudes

CHARLES LETESSIER AND VICTOR PURCELL

In the voluminous literature on traditional Chinese religion and cults some themes emerge very clearly. Purcell noted that 'a god is expected to render a return as the Roman gods were', and a large element of polytheism, expressed in the worship and celebration of many deities, with specific roles, had been added to the more abstract Confucian, Taoist, and Buddhist beliefs. As the reform movement in China, early in this century, spread to overseas Chinese communities, the reformers began to press for what they regarded as a purification of Chinese religious practice. But there was also resistance among the more traditional worshippers. The passages which follow have been selected to illustrate these trends.

The Temple of Sen Ta, Patron of Kuala Lumpur (1893)

A little while afterwards ... a temple bearing his name was forthwith raised by the exertions of Captain Ah Loi. This edifice occupies the angle formed by the junction of High Street and Pudoh Street, not far from the market....

In order to consult him recourse must be had to Thung Cen.

This Thung Cen is the medium through which the spirit has chosen to be manifested under the appearance of 'temporary possession'. This possession is shewn by the insensibility of Thung Cen to pain. For example, sticks of incense are applied to his ears without his evincing any signs of suffering. When he is to enter into this state of possession ('Kong Thung'), which he will do for the small sum of 50 cents from any private individual who invites him, he has himself memsmerised before the altar of the spirit by two familiars, who pass in front of him gold and silver papers lit from wax tapers, whilst he rests his head between his two hands, and presses it vigorously with two pieces of paper in the vicinity of the temples.

He is thus bent a little forward, and in this attitude he awaits the entrance of the spirit. All at once he gives a sharp cry like the cry of an owl and withdraws his hands from his forehead, while he works his head to and fro with groans, and his face becomes distorted like that of some madman or congenital idiot. His voice, articulating unintelligible sounds, recalls the noise made by certain big jungle birds. His assistants alone are able to interpret this singular language, which they note in writing when it relates to a medical prescription.

The End of the Chingay Procession? (1906–37)

Mr Ong Hwe Ghee was the chief speaker, and, arguing against the continuance of the procession, he said:

'The object of the temple is to commemorate the deeds of the departed great ones, to exalt their virtues, and to record appreciation of their services, and not to serve as a place for prayers and the asking for favours. Ignorant people subvert this evident policy with the vulgar notion that worshipping the gods will bring good luck and sacrifices will avert calamities. To make matters worse, busybodies add to the folly of the ignorant by introducing prancing lions and paper dragons—in fact, the Chingay procession—in order to please the gods. Now, even in China such processions are prohibited by law, although when the country is enjoying peace and prosperity the mandarins do not interfere.... As for respecting the gods, reverence, accompanied with the burning of incense, is enough. What need is there to belabour the people and waste money in order to compete in the vulgar shows in which unfortunate women are hired, and dressed up in gorgeous

Gateway of the Kek Lok Si Temple, the largest in Penang, from *Malaysia and Indo-China: Information for Visitors*. . ., Thos. Cook & Son, Ltd., Singapore, 1926.

style and paraded through the streets to be seen by all nations? This sort of show I consider to be an insult to the gods, and I tell you that if the gods have any sense of honour, you must be thankful if they do not curse you for thinking them capable of enjoying such rubbish.'

After further discussion the meeting unanimously agreed to the abolition of the procession and this was confirmed by a large meeting a week later.

This, however, is denied by Penang Chinese of to-day. Mr Gee Kok Weng says that a Chingay procession was held sometime in October 1928. In April 1935 another Chingay was held to celebrate the Silver Jubilee of His Majesty King George V. Again, in 1937, another Chingay was held to celebrate the Coronation. The Chinese, says Mr Gee Kok Weng, then took the opportunity to celebrate also in honour of the Goddess of Mercy.

The present writer witnessed both the 1928 and 1937 processions, which were on an elaborate scale, but did not realize that they were a revival of the Chingay whose demise was announced in 1906!

Fr Charles Letessier, unsigned note entitled 'Si Sen Ta—A Chinese Apotheosis', *Selangor Journal*, Vol. 1, pp. 321– 2; Victor Purcell, *The Chinese in Malaya*, Oxford University Press, London, 1948, pp. 125–6, quoting a report of a meeting on 16 December 1906 at the Hokkien Temple, with additional comments by Purcell.

Schools and Hospitals

52
Hospital Horrors

CUTHBERT HARRISON

The introduction of Western medicine to Malaya in the latter part of the nineteenth century met with a mixed response. In a practical fashion the local communities recognized that the new medicines and dressings were efficacious. But there was fear of surgery, vaccination, and other apparently injurious treatments, and above all misgivings about going into hospital as an in-patient vulnerable to the starvation diets and barbarities which European doctors were believed to enjoy inflicting.

THE hospitals of Malaya are all owned by the State, the few Chinese-run hospitals being more in the nature of homes than hospitals as we know them. A description of any native hospital will serve very well for those at Taiping, Ipoh, Kuala Lumpur, Seremban and many other smaller places, for all are run upon uniform lines. The first point to note is that the hospital stands inside a very high ring fence of wire so closely knitted that it is not possible to pass the hand through. This fence is not to keep the patients in, for every one is free to leave whenever he will, but it is intended to prevent well-meaning friends from passing food through to the patients from outside. Here is the first point of cleavage between the Occidental's and the Oriental's idea of medical treatment. The Oriental is still at that stage of thought on medical subjects which is found amongst the lower and more ignorant classes of Europe. He believes in food and plenty of it at all times. The very idea of dieting a patient is strange to him. That a man half-dead of dysentery should not be allowed to eat curry and rice seems cruelty to him and his friends. If you walk through a hospital and ask 'any complaints?' some one, some Tamil or Chinese, is certain to hold up his hand and state a grievance. You stop and listen, expecting perhaps some complaint of harsh treatment of a patient by a native dresser. But no—your grumbling patient only tells you 'They don't give enough to eat—nothing but slops—no rice, no curry. My inside is empty, empty, Tuan!' With a smile you turn to the English surgeon in charge who tells you 'Yes, dysentery case. He had a relapse about a week ago. We couldn't account for it. At last we discovered that his wife had thrown some curry and rice, wrapped in a leaf, over the fence. He had eaten it. Result, relapse, and he nearly died.' One admires the devotion of such wives, but wishes they knew more of the effects of curry and rice upon the dysenteric human interior. But, after all, is it wonderful that she thought her husband was being slowly starved to death, for even more ridiculous superstitions about European medical treatment are very current. The Malays, for instance, most firmly and fully believe that if a patient is admitted to hospital and is not cured within a few days, the white doctor poisons him off. Purely fantastical though this belief is to us, it is yet based on a cross-eyed logic which convinces the Malay mind. The idea arises thus: As all the races in the country have a great belief in European *drugs* which they can take as outside patients and prove in their own homes, so also they have a great horror of European *treatment* involving segregation in a hospital, and often surgery, 'cutting pieces off people'. Those two sides of it frighten the imagination. But to their frightened imaginations they further present the well-known fact, carefully acknowledged in all Government returns, that an enormous quantity of admissions to hospital die within twenty-four hours of admission. The Malay puts two and two together and to his own horrified satisfaction makes five of them. Says he, 'It is plain. All men know it. See how many go in and how few come out. There is a reason for this. The reason is that if a doctor sees he cannot cure a man he is bitterly ashamed. He says to himself, "this person shall not linger here to bring shame upon the art and practice of medicine. Better dead!" So he poisons him and afterwards they deliver the corpse to his friends and to the Kathi, who bury it. That is the way of it.' That, of course, is precisely not the way of it. The true way of it is simpler than that and not so titillatingly horrible, being merely that people who will only resort to a hospital when they are at the last gasp will naturally die in the hospital, as indeed they would have died outside. Asiatic patients cannot bring themselves to enter a hospital until they have exhausted every native treatment. They are really embarked on the last long journey before they are taken to hospital, and it follows thus of a certainty that deaths follow admissions very rapidly. The doctors trust to time, education and demonstration to kill these ideas, but though an impression is already made, Asiatics love these beliefs and cling to them with misplaced enthusiasm very galling to the medical profession. After all, they are no more ridiculous than many a superstition still current among the peasantry of Europe.

The buildings in the hospitals are all of a similar type, and consist of long airy wards, floored in cement, and lined with rows and rows of plank and trestle beds. The only race in this country which makes a practice of sleeping upon a soft mattress is the European, and he does so for the excellent reason that it is the custom in Europe. All the other races rest upon plank beds on which a grass-woven mat is laid, the Malays even laying the mat upon the floor at times. So the beds in the hospital are all planks, and each is provided with a red blanket and a wooden pillow. The pillow is wooden for the same reason as the bed is of plank, the patients being accustomed to hard wooden pillows in their own homes. If you gave them a soft pillow they would complain. That curious cement tank in the grounds with a worm-screw and press arrangement is the place where the bed-boards are periodically soaked in disinfectants, for pauper patients are much infested with bugs and other creepy crawlies. The various diseases are kept apart as much as possible, and you will usually find a ward for beri-beri, a ward for dysentery, another for phthisis, and another for malaria, and perhaps others as well. Everything is clean and neat and, if you can put up with the smell of disinfectant and the sad incidents of illness, a hospital in Malaya is well worth visiting.

In or near the headquarters hospital in each State is the lunatic asylum, but shortly all lunatics will be taken care of at a central asylum at Tanjong Rambutan. The most prevalent form of lunacy in Malaya is melancholia, a quiet form of insanity which permits of the patients being kept together in association and employed in useful spade labour, either in or near the hospital, an occupation to which they have all long been accustomed before their mental powers failed. Many a madman has had to thank this daily round and common task of digging for his recovery.

Somewhere near the hospital and the lunatic asylum will be the leper ward. Amongst the many benefits which the British have brought to Malaya we cannot, alas, yet reckon a cure for that horrible disease of leprosy. For people affected with it little or nothing can be done for certain, but as they are regarded as a danger to their fellows they are segregated, some in leper wards on the mainland and some on the leper islands on the west coast of the Peninsula. The Malays have a horror of leprosy, and use various euphemistic expressions even to describe it. The duty of capturing lepers is intensely repugnant to the native headmen and the native police, and no one would ever willingly come forward and report a leper for deportation. Therefore at intervals the District Officers assemble their penghulus [headmen] and require of them each a report to be sent in by a definite date as to the whereabouts of each known Malay leper in the district. The penghulus then, each in his mukim [district], make cautious enquiries as to whether anyone has contracted the disease since last investigation, or whether any stranger suffering from it has entered the mukim. Such enquiries are of necessity cautious, for no one would willingly disclose the existence of leprosy in a father, a mother, a wife, may be, or a husband, since segregation is certain to follow if the medical examination confirms the native diagnosis. With a pitiful devotion all kinds of shifts are tried. The affected one will live amongst the family and all will run the well-known risk of contagion, or perhaps he or she will be sent to live in some hut in the jungle, far from the habitations of men, a fugitive and outcast, fed by someone's loving care, solitary, rotting steadily with cureless disease. It may be that, refusing to recognise the first symptoms, the sufferer will resort to the house of some native doctor, there to be slowly bled of any money he may have and slowly to watch that hideous development. Whatever evasion be practised, at last all will prove vain, for someone in the secret will either wish to curry favour with the penghulu by informing or will have a spite against the patient or the family. So at last the case is located and one early morning the penghulu and the police will attempt the capture. Advisedly we say attempt, for they do not always succeed. These poor creatures, clinging to their liberty, infected and infecting though it be, will often hide from the authorities and escape capture for long. Yet if they only would believe it their lot is, except for separation from their homes, far happier on a leper island than anywhere else.

C. W. Harrison, *Illustrated Guide to the Federated Malay States*, The Malay States Development Agency, London, 1911, pp. 124–30.

53
Latah

ARTHUR KEYSER AND RICHMOND WHEELER

Like amok (see Passage 16) the mental affliction known as latah, *observed among Malays in earlier times, has now almost disappeared. An eminent doctor, Sir James Galloway, who practised in Singapore at the beginning of this century, believed that the quietude of Malay village life exposed some innately vulnerable individuals to* latah *attacks if they were subject to sudden disturbance. L. Richmond Wheeler, in the second passage quoted here, draws on his experience in instructing Malay student teachers, and places more emphasis on the traditional deference of Malay villagers to the ruling class as a determining factor. In a noisier modern world of transistor radios, motor bikes, and social equality, those who might have suffered* latah *are presumably 'hardened' against it in their youth.*

MALAYS were liable to a malady which seemed quite peculiar to their race—it was known as 'Latah'. A man might have this affliction and yet pass through many years of his life undetected, until it was aroused by some sudden movement or sound. A man or woman with 'Latah' is impelled to imitate the motions or noises of others. They are often victimised by heartless practical jokers. I have seen the latter pretend to dive into a river, and the unfortunate 'Latah' man would actually plunge in, just as he was, fully dressed. I have seen others play crueller pranks by feigning to tear off their clothes, inducing the poor imitator immediately to discard his, or her, own. I was once riding when my horse neighed. An old woman coming along the path towards me suddenly threw up her hands and neighed in return, and, as she continued this conduct while prancing in front of the horse, it was with some difficulty that I was able to guide him past without hurting her.

My [police] Sergeant-Major, a very smart and dignified officer, was interviewing me on my verandah. My tame cockatoo, emitting some of its most horrible notes, hopped on to the floor. To my amazement, the visitor, oblivious of his position and his uniform, immediately shrieked in reply, and going on to his knees proceeded to hop about the floor in faithful imitation of the bird. He was able to recover himself by an extreme effort of self-control....

Natives have been seen sitting on the outskirts of the jungle bowing repeatedly to the swaying branches of trees moved to and fro by the breeze. Bound by some irresistible spell they could not move and come away.

* * *

All observers have noticed the difficulty of getting a proper reply from Malays when they are addressed suddenly, and the nervousness that they manifest when they confront rajas and Europeans, even those from whom they have nothing to fear and whom they know well. These symptoms vary, of course, in individuals, there being some with strong personalities who bear themselves with dignity and self-possession even before strangers. But, to take a personal instance, though no difficulty arose with the best students, with a large number of the lads at the Sultan Idris Training College I have found the usual method of asking quick questions in order to test progress often failed utterly, because the candidate was so hopelessly upset, and would again and again fail to avoid the most obvious errors under oral examination when his written work would be quite reasonably accurate. It is, for instance, absurdly easy to get a student of seventeen or so to reply that the sun sets in the east or the north, or to make a similar blunder over simple facts of general knowledge, such as the name of the local river or the best-known food crops. Similarly, though there are hardened souls who will brazen things out, the majority, when hauled up before authority for some small breach of conduct, are in such a confused and cowed condition that it seems cruelty to do more than hint at the possibility of improvement. And the number on whom this mild correction does not have the desired effect is very small; it consists almost entirely of the few who are really unfit, mentally or physically or both, for the standard required.

A. L. Keyser, *Trifles and Travels*, John Murray, London, 1923, pp. 94–5; L. Richmond Wheeler, *The Modern Malay*, Allen & Unwin, London, 1928, pp. 220–1.

54
A School at Kuala Pilah

GERWYN LEWIS

Gerwyn Lewis began his career in the Malayan Education Service in 1938. After suffering the hardships of a POW on the Burma–Siam railway he returned in 1946 to take up the post of headmaster of the secondary school at Kuala Pilah in the heart of Negri Sembilan. He completed his distinguished career with a long spell (1956–62) as headmaster of the Victoria Institution, Kuala Lumpur.

WE learnt of the terrible sufferings of the people of Kuala Pilah during the Japanese occupation, how the Tuanku Muhammad School had been used as the headquarters of the Japanese *Kempetai* (or secret police), how hundreds of people had been tortured at the school, and how over three hundred mainly Chinese victims lay buried in the small rubber plantation at the back of the school.

When I arrived at the school I found it in a lovely location, overlooking a well kept *padang* and surrounded by shade trees. One of its former teachers (Mr Francis) had successfully re-opened it, but it was operating under many difficulties, especially the lack of books. So my earliest efforts were directed at acquiring textbooks and building up a school library.

One day when I took one of the senior classes, I recognized

one of the Malay mature students. He had worked like myself on the Burma–Thailand railway as a medical dresser. His name was Mohd. Sharif bin Ishak. He wrote an article in the first issue of the school magazine about it, and described his experiences as a dresser in one of the Japanese jungle 'hospitals' as follows: 'I have never seen such suffering of human life as I did in that hospital which could be called a living hell.' But we never discussed our mutual experiences, for we both wished to forget that nightmare.

There was keen interest in education, especially from the kampong Malays, which was a welcome development. In fact they were so enthusiastic that many were forging the dates of birth on the birth certificates of their children so as to qualify for entry into the Tuanku Muhammad School. One parent even offered me an envelope which, when I opened it, contained a bundle of ten-dollar bills, perhaps one hundred dollars in all—a lot of money in the 1940s. So I gave them all a lecture on the evils of forgery and bribery and told them not to do it again! Some were a bit ashamed of themselves, but all were puzzled how I knew they had tampered with their birth certificates, as the forgeries had been beautifully executed. They had overlooked the fact that the serial number at the top of the birth certificate, also included the year of issue as a part of it!

As many of the pupils were still suffering from several years of malnutrition, free Klim milk was issued to all pupils every day. I found a cheap source of Klim milk for my own use in the town at a shop run by a shopkeeper with whom I was on friendly terms. I thought I was getting my milk cheap because of our friendship, until Stewart Angus the District Officer told me one day that someone had stolen a lot of his Klim milk that he had stored in a building in his garden!

There was also a great shortage of all kinds of consumer goods, and in an attempt to alleviate this shortage fairly, I was supplied with bales of cloth for distribution amongst the pupils. As for myself I had no car, for none were available, and often walked to school or cadged a lift from Martin Read who was allowed a car because he was the Medical Officer, or Miss Bunty Coupland, my so-called 'European Mistress' who also had a car for some of her special duties.

We were, of course, short of many other things, especially cutlery for special occasions. Consequently our Malay servants frequently borrowed each other's employers' cutlery. Thus one evening at a formal dinner party at the doctor's house, I unthinkingly admired his cutlery and remarked that we had a somewhat similar set. It was only later I discovered that it was in fact our cutlery, and that Puteh our Malay servant had lent it to the doctor's house. *Pinjam sahaja* (borrowed only) explained Puteh in her loud infectious laugh!

Almost a year after my arrival at Kuala Pilah, Lyn who had so patiently waited for me for such a long time during the war, was allowed to come out and join me. I went to meet her at Singapore in March 1947 and we travelled from Singapore to Seremban by train. When we had gone about half way, Lyn asked me why our Tamil Station Master travelled with us on the train. I explained that the Tamil Station Masters she had seen at our various halts were not the same one, but different ones who looked the same because they all came from Ceylon!

We arrived at Kuala Pilah in a thunderstorm. She described it in her diary after her arrival at Kuala Pilah as follows: 'We eventually set out on the 25 mile drive to Kuala Pilah and home. There is a range of hills between us and Seremban and we had to cross them. The road wound between jungle covered hills, the sun was setting and the view was wonderful. It got dark before we arrived and the rain poured down in sheets and there was tremendous thunder and lightning to herald my arrival. I can't remember much about the journey, I was too excited.

At the door was Puteh—she is now an old friend though our conversation is extremely limited, and the old gardener and his wife. Puteh gave me a beautiful bunch of flowers, white wax-like flowers, the petals white at the edges but yellow at the centre and giving out an exquisite perfume. They grow on a tree in the garden and I can't remember the name. Puteh is charming and has a lovely sense of humour and a very infectious laugh.'

Lyn loved Kuala Pilah and was fascinated by her new environment, by the friendly people and the variety of customs. She also made friends very easily, including one of the royal family at Sri Menanti. She was also asked by the District Officer if she would like to help, on a voluntary basis, with the social welfare programme of the kampong Malays and readily agreed. In this connection she visited many kampongs, some of them very much off the beaten track. One of these, not far from Kuala Pilah had suffered a traumatic experience. Apparently when the Japanese were defeated and the Chinese Communists came out of the jungle, there was trouble between the Malays and the Communists at this kampong, as a result of which some of the Malays took the law into their own hands and killed several of the Communists. British justice being what it is, with its insistence on the rule of law, the Malays involved were had up for murder and sent to jail at Kuala Lumpur. Unfortunately the poor wives of those kampong Malays had no idea how long their husbands had been sentenced to jail and were in a great state of economic distress. Lyn and the Welfare Nurse eventually sorted things out for them, found out how long their husbands would be in jail, and arranged for them to get some economic assistance from the Social Welfare Department.

Our house was located on the side of the hill at Kuala Pilah and overlooked a Malay kampong. Consequently the kampong noises and the sounds of popular Malay songs such as the delightful one called *Rasa Sayang*, would often drift up towards us providing a romantic background to our home. We were also located a few metres from the jungle and on moonlight nights wild pigs would often come and feed off the nuts which fell off a nearby palm tree in the garden.

Puteh was full of superstitious beliefs and folklore. One moonlight night she came hurrying upstairs where Lyn and I were having a drink with Stewart Angus, the District Officer, and announced that she had just seen a tiger. So I rushed to an upstairs window with a strong torch, and there below me walking along the road leading away from the house I saw not a tiger but a large dog. When I informed Puteh of this fact, she became even more distressed, and said it confirmed her fear. The tiger she informed me was *Datuk Gaung*; who when necessary could change himself into a dog!

G. E. D. Lewis, *Out East in the Malay Peninsula*, Penerbit Fajar Bakti, Petaling Jaya, 1991, pp. 139–42.

55
A Schoolmaster's Life is Not a Happy One

ANTHONY BURGESS

Before he became one of the leading English novelists, Anthony Burgess worked from 1954 to 1960 in the Education Service in Malaya and then in Brunei. This experience provided the background for a sequence of three novels set in those countries. This passage comes from the second of the trilogy. Burgess's immense talent combines acute observation with derisive caricature. Some of the details in this passage, such as Ah Wing's eccentric diet, may fall into the latter category.

THE thermometer in Crabbe's office read one hundred and six degrees.

It was not really the office, it was a book-store. The real office had become a class-room, housing the twelfth stream of the third standard, and soon—the *kampongs* milling out children in strict Malthusian geometrical progression—the store itself would become yet another class-room. Then Crabbe would have to take his telephone and typewriter into the lavatory. He went to the lavatory now, without typewriter or telephone, to pull off his shirt, wipe down his body with a towel already damp, and drink thirstily of the tap-water, brownish and sun-warm. For a fortnight now he had intended to buy a large thermos jug and a table-fan. His prodigious dryness was due to the heat, too much smoking, and to a great salty breakfast.

The excessive smoking was the fault of Haji Ali College; the Pantagruelian breakfast was Ah Wing's regular notion of what an expatriate officer should eat before going off to work. Crabbe had consumed grape-fruit, iced papaya, porridge, kippers, eggs and bacon with sausages and a mutton chop, and toast and honey. At least, these things had been set before him, and Ah Wing had watched intently from the kitchen door. Crabbe now saw that he would have to beg Ah Wing to go and work for Talbot, a man who would meet his challenge gladly, might even call for second helpings and more bread. But perhaps Ah Wing would cunningly recognize Talbot's greed as pathological and despise him more than he despised Crabbe, deliberately burning his steaks and under-boiling his potatoes. Somewhere in Ah Wing's past was a frock-coated whiskered law-bringer who had established the pattern of square meals and substantial heavers. Perhaps Ah Wing was to be seen on some historical photograph of the eighteen-seventies, grinning behind a solid row of thick-limbed pioneers, all of whom had given their names to ports, hills, and city streets. Certainly, in the gravy soups, turbot, hare, roast saddles, cabinet puddings, boiled eggs at tea-time and bread and butter and meat paste with the morning tray, one tasted one's own decadence: a tradition had been preserved in order to humiliate. Perhaps it really was time the British limped out of Malaya.

Ah Wing was a fantastic model of Chinese conservatism. He had not at first been willing to recognize that Crabbe was a married man and had set only one place at table. At length, grudgingly, he had obeyed his master's sign-language. And habits of long repatriated officers seemed fossilized in certain rituals: a large bottle of beer was brought to the bedside on the morning of the Sabbath; twice Ah Wing had entered Crabbe's unlocked bathroom and started to scrub his back; once Fenella had been rudely shaken from her dawn sleep and told, in rough gestures, to go.

But Ah Wing's private life—in so far as it showed chinks of light in Crabbe's kitchen-dealings with him—exhibited a much more formidable conservatism, dizzying Crabbe with vistas of ancient China. He had once caught Ah Wing eating a live mouse. A day later he was proposing to send a black cat after it (black cats were said to be tastier than tabbies). Then there was Ah Wing's store of medicines—tiger's teeth in vinegar, a large lizard in brandy, compounds of lead and horrible egg-nogs. Crabbe discovered that his cook had a great reputation with the local Malays with whom he did a roaring trade in—eventually lethal—aphrodisiacs. The village *bomoh* was jealous of Ah Wing and called him an infidel. And indeed Ah Wing's religion, though not quite animistic, was too complex and obscure for enquiry.

The local Malays were a problem. They squattted on Crabbe's veranda every night, waiting for tales of distant lands and Western marvels. Crabbe had now established the routine of reading them love *pantuns*—mysterious four-line poems he

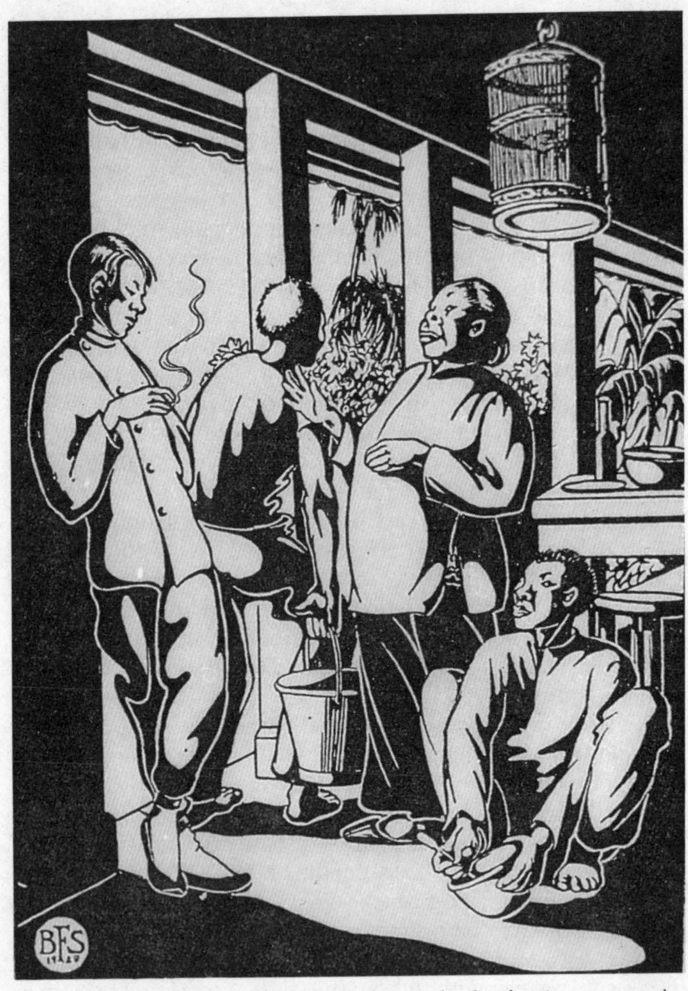

Some of the domestic staff—from left to right the 'boy', water carrier (*tukang ayer*), maid (*amah*) and cook, on the back verandah, from Ashley Gibson, *The Malay Peninsula and Archipelago*, J. M. Dent & Sons, Ltd., London, 1928.

had found in a Malay anthology—and discovered that a few verses went a long way, producing ecstatic cries, grunts of deep approval, profound nods and writhing of bodies. Here at least was a healthy literary tradition. It was Fenella who suffered most. The Malays were fascinated by her fair hair, and children were brought along to clutch it stickily, as against the King's Evil. The women asked her about her underwear, begged for discarded brassieres, and went round the lounge, handling the ornaments and wanting to know how much they cost. It had taken Crabbe some time to become accustomed to Malay elders squatting on the dining-room floor and using the second bedroom for reciting their prayers. His predecessor had been decidedly too chummy. And Fenella was still scared of taking a bath, for a spirit of sincere inquiry sent serious Malay youths to the bathroom window to find out if white women differed materially from brown ones. It was not easy, and Fenella talked more and more of going back to England.

Anthony Burgess, *The Enemy in the Blanket*, Heinemann, London, 1956; reprinted by Penguin Books, 1972, pp. 271–3.

56
Teaching at the University of Malaya

JAMES KIRKUP

To James Kirkup (Passage 29) the Majestic Hotel was more congenial than his new workplace—the University of Malaya—but he provides an equally vivid picture of his students and colleagues.

MY first lecture, on *The Portrait of a Lady*. The weather that morning: a slogging, wet heat and an almost blinding haze. I went into the super-modern, air-conditioned University library to cool off a little before the lecture and put on a few drops of Christian Dior Eau de Cologne Rafraîchissante. It wasn't very.

The lecture-theatre was large, air-conditioned, with a film projector in a glass cabin way high up at the back, nearly touching the ceiling. The seats were on such a steep rake and so close to the rostrum that I felt I was standing at the bottom of a well. Green blackboards, all scrawled over: I had to clean their vast expanses before I started; it was like swimming the Channel in a snowstorm. So I was not in a very good mood when I turned to face my students for the first time. They were first-years—I had arrived in the middle of the second term of the academic year—about fifty of them, mainly females wearing saris, Chinese pyjamas, cheongsams, sarongs or Western-style clothes. Some of them were extremely pretty. Only five men; three of them, wearing glasses, were Chinese: there was one Indian and one Malay, a very good-looking and intelligent-eyed one sitting far apart from the rest. Curiously enough, I never saw him again: he must have been a wild one.

They understand and speak English, and presumably write it too, with remarkable fluency and accuracy. They get my jokes with the same quickness as the Japanese. They also love to be given 'outlines' and 'projects' and lists of 'themes' and periods and so on, which is very tiresome....

Most of the Europeans are vulgar without, unfortunately, being coarse, and their society is one of the most hag-ridden and commercially-minded in the East, besides being philistine, anti-intellectual, anti-creative, anti-egalitarian and plain snobbish. Behind the University's brash modern façade there is an almost total lack of intellectual and cultural life, a fact frequently referred to by bored and frustrated students. I felt at the time that besides having come to the world's most boring capital I had also been trapped in the world's dullest university, where most of the staff's free time is spent exhibiting feeble academic wit at endless committee meetings, staff association meetings, faculty meetings. There was no Common Room, no place where the members of the teaching staff could meet each other informally and sociably. I very soon gave up going to their gatherings where one simply sat and listened to people giving displays of academic and committee jargon, instead of meeting them. Stay here for *three years?*

James Kirkup, *Tropic Temper: A Memoir of Malaya*, Collins, London, 1963, pp. 45–8.

Plantations and Planters

57
Opening a Rubber Estate in 1913

HARRY PIPER

Harry Piper had served in the Boer War and tried other occupations before coming to Malaya, one of the many who sought a new career in rubber planting. After learning his trade as an assistant on another estate he obtained (in 1913) the post of manager on what became the Sua Betong estate in the Port Dickson District of Negri Sembilan. The pioneer planters had to turn their hands to whatever tasks were most pressing—in this case the movement of heavy machinery for installation in the estate factory. He goes on to describe a close encounter with a tiger in rather unusual circumstances. Harry Piper, a formidable personality, rose to become the senior manager of one of the largest plantation groups in Malaya. He died, after a short retirement, in 1956.

THE difficulties met and overcome in transporting this engine from the 16 mile Government Road (Ayer Kuning Corner) to the Sua Betong Proper Division, a distance of 10 miles of bullock cart track running through 10,000 acres of *lalang* [coarse grass] land, are worth relating.

The fly wheel and engine base weighing two tons and one ton respectively were the difficult parts. The fly wheel arrived at Ayer Kuning Corner, i.e where the bullock cart track commenced, on a low lorry with wheels about 2 feet in diameter which was pulled by six pairs of oxen pulling six carts. This was in charge of Mr Whitehead, Messrs Guthrie & Co's engineer. Mr Whitehead had recently arrived from England and was in charge of the transporting and erection of the engine. Before he had seen a mile of the road over which the engine had to pass, he said it was an impossibility and warned us to take no further actions until he had consulted his employers.

Realising there was no alternative route over which to take the engine, the writer and his assistants, Messrs Topliss and Johnstone, decided to try and transport it and commenced immediately the engineer had left....

For the first half mile the track was low lying and very sandy and we had to lay down planks to prevent the wheels of the lorry from sinking into the sand. Hair pin corners, of which there were several, and the bridge spanning the Linggi River, were the other main obstacles. When going up hill, there were at times six pairs of oxen in six carts pulling the lorry.

The 40-foot bridge, spanning the river, with a hair pin bend at the end, was not built to take anything like the weight of the fly wheel, lorry and oxen. I remember how we quaked with fear, on seeing the bridge sway, lest it should collapse and deposit the fly wheel in the bed of the river some 20 or 30 feet below. To add to our worries, the engineer in charge had returned, placed full responsibility on my shoulders and went his way. The engine—complete—was on the site of its erection within one week.

The 10 miles of bullock track, running through thousands of acres of *lalang* land, which connected up the three divisions and gave the estate its only outlet to the Government Road 16 miles from Seremban, and a somewhat similar distance to Port Dickson, was improved and kept in repair by the filling in of pot-holes, and making culverts and side drains where necessary.

With the permission of the Directors of the Port Dickson Rubber Company, whose property joined the Sua Betong Proper Division, a gravelled road was constructed at the expense of the Sua Betong Rubber Company through their, the Port Dickson Rubber Company's Sirusa Division, which joined up with their Port Dickson estate road, over which we were given a right of way. As this road met the Government Road at the third mile from Port Dickson, it gave us a road through to Port Dickson.

My first experience with [a] tiger was shortly after my arrival to take up my duties on the Sungei Ujong Division. One evening after dinner I had occasion to pay a visit to the lavatory, which was situate some twenty five yards distant from and behind the servants quarters. The lavatory was a very primitive affair, a hole in the ground, with a hut built round it 4 feet x 4 feet with a corrugated iron roof. It rested on top of the ground. When I had been inside a minute or so, fortunately with the door closed, I was conscious of what appeared to be a large animal prowling round the building, sniffing and breathing heavily. Too scared to open the door and make a bolt for the bungalow, having no kind of weapon to hand, I had to be content with shouting at the top of my voice, with a view to attracting my Chinese servants' attention. After the longest hour I have ever known, I heard them coming towards me, beating an empty kerosine tin. They had just returned from the Chinese Kongsi, some two hundred yards away, and had only just heard my shouting.

During the time I was a prisoner the fact dawned on me that should the animal lean hard up against the building I was in, it would probably tumble over, in which case I had decided, if possible, to disappear down the hole in the ground, in preference to becoming a meal for what I had made up my mind was a tiger. This surmise proved to be correct when, the next morning, numerous large tiger pad marks were to be seen round the outside of the building.

Harry Piper, unpublished memoirs of his planting career written after his retirement in 1949.

58
A Warm Welcome

LEOPOLD AINSWORTH

Ainsworth, like Piper (Passage 57 above), was one of the early rubber planters. But estate cultivation ('planting') of other crops, such as coffee, had established a tradition that beginners (called 'creepers') were often given a rough apprenticeship in hard living conditions, to prepare them for a demanding career—and to weed out any who were not tough enough for it.

Ainsworth here describes his arrival, as a new assistant recruited in London, on a rubber estate in south Kedah. His journey from Penang had been an ordeal, full of difficulties. But it earned him no sympathy from his new manager. In time Ainsworth came to like and respect the autocratic and eccentric manager (A. W. Davidson), though he later moved on to pursue his career elsewhere.

THE cart at last drew up outside a bungalow, and after a few minutes' wait a short, thick-set white man came out, and as he answered to the description of the manager which I had been given, I introduced myself to him as the new 'creeper'.

I was tired out, ravenous and thoroughly dazed after my strange journey, but my heart sank when he merely looked me up and down with every symptom of disgust and bad temper, and asked me curtly why I had arrived at this ungodly hour. I told him of the delay at the river's edge, the non-arrival of the bullock cart, and the subsequent discovery of the senseless driver at the Chinese mill. The news seemed to lash him into a perfect frenzy of uncontrolled rage, for without saying another word to me he turned upon the driver with a torrent of unmistakable abuse, and proceeded to give him a most terrific thrashing, which struck me at the time as being a queer way of settling the fare. After rousing the Chinese cook by a further torrent of fury, to come out and give a hand with my luggage, we went into the bungalow and waited there, eyeing each other while it was being brought in, till after a few minutes of silent contemplation he very abruptly remarked, 'I suppose you want something to eat.' I replied that I had not had a meal since sunrise and was starving, but before I could enlarge upon the subject further he bellowed to the cook, 'Bawa makan sama tuan' (which meant, 'Bring food for the master'), and left me to wait for it, saying that he himself was going back to bed. He pointed to a door upstairs where I should find my luggage, and told me that I should be called at 5.30 a.m.

Without saying good night or even troubling to ask me about my journey he stumped off, leaving me wondering what sort of a man I had run up against, and how we should get on together if we had to share the same bungalow. He was certainly the toughest man I had ever come across, and something quite new to me.

After a dreary wait of twenty minutes I had my first rough meal, which at the time I thought far and away the most disgusting collection of food that had ever been brought together at one sitting. Soup appeared first; tinned, of course, but with the additional nourishment supplied by the corpses of numerous small ants which formed a thin crust on its surface. Next came tinned soft roes on toast, tasteless, uninteresting, and badly fried in coconut oil. These were followed by some form of tinned beef, exquisitely tough and stringy and flavoured with a sharp metallic tang. There was no bread, its place being taken by large dry biscuits of the dog variety, very hard and tasting of cardboard and straw, whilst the butter wallowed, semi-molten, in a tin, and smelt, looked and tasted beyond my powers to describe.

The last course consisted of banana fritters, which looked good, and undoubtedly would have been but for the fact that the flavour of the bananas was completely lost in the overpowering taste of the coconut oil in which they were fried. The meal terminated with coffee, which was equally shattering, though it was not until later that I discovered its horrible flavour to be due to the cook's invariable habit of using one of the manager's socks as a strainer. Nothing appeared to his mind quite so effective for the purpose as this, and he was modestly proud of his discovery, so that no amount of punishment or threats could wean him from the use of it, whilst confiscation of the offending article merely meant that ere long another would be stolen and conveniently hidden for secret use in the future.

Being ravenous, I picked my way through the meal, dodging as best I could the ants which garnished every course of it, till at last it was finished, and I went up to my room to unpack a few things and turn in for the remaining hours before half past five.

My bedroom was just a bare boarded room containing a bed and a large cupboard, and lit by a small oil-lamp on the wall, which gave a miserably dreary light and gloomy appearance to everything; however, I was soon in bed under the mosquito curtains, but found sleep difficult. The mattress and pillow felt terribly damp and smelt strongly of mildew: blankets and sheets there were none, nor were these missed in the oppressive damp heat of the room, and I found myself perspiring even without them. On the bed was a large bolster-like object called, as I subsequently discovered, a 'Dutch wife', and this I found to be a great comfort when used in the proper way, which is to place it between one's legs in order to keep them apart to allow air to pass between them, for the humid heat is so great that if one's legs touch each other they perspire profusely. . . .

It seemed as if I had scarcely shut my eyes before there was a terrific banging on the door, and I woke to the fact that it was half past five in the morning and time to get up; but I lay in bed for a few minutes longer, till the battering on the door was repeated, and an angry bellowing voice roared, 'Get up, you lazy b—, how many times do you expect to be called?'

I was out of bed like a shot from a gun and in an instant was at the door, for after such a night my nerves were all on edge and my temper short. This particular abuse was offensive, and in a white heat of rage I flung open the door and, finding the manager standing outside, threatened to punch his head unless he apologised for what he had said. He seemed somewhat taken aback and surprised at my fury, and assured me that no offence was meant, and that the expression he had used was almost a term of endearment in these parts. I accepted this explanation, and decided that I should have to learn this new language.

Leopold Ainsworth, *The Confessions of a Planter in Malaya*, H. F. & G. Witherby, London, 1933, pp. 39–42.

59
Planters of the Old Days

JOHN HOCKIN

Stories about the first generation of rubber planters, many of them Scotsmen, were cherished and provided the humour for after-dinner speeches, etc. No doubt they gained somewhat in the telling.

THERE were lots of Scotsmen among the pioneer planters; there has always been a strong strain from North of the Tweed on estates in Malaya. The early Scotsmen were frugal in their habits, shrewd business men and prodigiously hard workers, regardless of the climate. They took to the life more readily than the majority of Englishmen.

Lots of tales were told about them. There was one old Scotsman, for instance, who consumed his regular bottle of whisky a day but who would solemnly lecture the assistants he appointed at £50 a year on the evils of alcohol. Water was much better for them than wine, spirits or beer, and the water on the estate was 'verra gude'.

There was another Scotsman with a Scottish wife who took a 'creeper', that is, a planting apprentice who pays a premium to be taught the job. In addition he pays so much a month for board and lodging in the planter's bungalow, and it follows, of course, that the less the 'creeper' costs to keep, the more profit is made out of him.

One day the wife was heard calling frantically.

'Keith, Keith!'

'Whatever's the matter, my dear?'

'Keith, quick—the creeper's awa' to the bathroom with the scented soap.'

The husband, it need hardly be added, died a very wealthy man.

Then there was a Scotsman, not a hard worker this time, who was once asked by the Visiting Agent, who came to inspect his estate, whether he had seen the weeding in a certain field.

'Na, na, sir,' was the astonishing reply. 'The assistant does all that sort of thing. I sit in the house and think, and give directions.'

Even 20 years ago there were still a few of the old school of Scotsmen in active control of estates. One old man I knew started off in Ceylon in 1872, went to Malaya some 20 years later, made a pile in the rubber boom, had lost it all a few years later in injudicious speculations in rubber shares, and finally came to finish his days on a Ceylon tea estate.

He remained a gambler, and also a bachelor (he never allowed a woman inside his bungalow and this in the 1920s), and still delighted in poker parties until the 'sma' wee hours'. If he were winning, the old man would start yawning vigorously not long after midnight, but if he happened to be down, then he would take it as a deadly insult if anyone suggested leaving before the tom-tom calling the coolies to muster at 6 in the morning....

There were a few brave women who were willing to face the privations and loneliness of estate life. There was one famous planter's wife in the old days who not only managed her husband with complete success but also ran the estate as well, not entirely without friction, it must be admitted, with the assistant managers. She made a regular daily round of the estate on horseback with her husband trailing along behind, and woe betide the young planter who had not carried out her orders to her complete satisfaction!

J. Hockin, 'Planters of the Old Days', *British Malaya* (monthly magazine), April 1945, p. 137.

60
A Planter's Life Today (1948–9)

ANON

In the communist rising, known as 'the Emergency', the objective was to take control of the outlying areas, and from those bases move towards the centre—the strategy of Mao Tse Tung in China. In Malaya the attack was eventually halted and crushed in the outlying areas, but only after years of struggle between the insurgents and the security forces. The 'state of

Emergency' (on which see also Passages 99–100 below) began in 1948 and was officially ended in 1960. In the early years, when the initiative was with the insurgents, the staff of estates and mines and the population of the rural villages were exposed to constant attack. The author was an estate manager in Selangor.

THE distance from my bungalow to the village is some three miles by road, but as the crow flies, about one mile. All was peaceful on the evening of September 17th—we had been listening to the radio—when bursts of Sten gun fire rent the air. This sounded unpleasantly near and we all manned our defences without delay.... Nothing eventuated in the bungalow area but we got the story of what had happened the next morning. 30 armed bandits had suddenly arrived in the village. They closed the level crossing gates and posted two sentries to see that they remained closed. The remainder then split into two parties. One went to an eating stall, called for the proprietor and shot him dead. The other party proceeded to a gambling saloon, called for the proprietor and shot him dead. Both parties, leaving their victims' corpses in their respective shops, returned to the level crossing, collected their sentries and withdrew. A Police force eventually arrived but no contacts were made with the enemy.

Nights of being disturbed every hour by the telephone check-calls, by false alarms and by having to make routine inspections of the Specials began to have its effect, and both I and my assistant began to feel washed out. I told the assistant to arrange to get away for a week-end's rest. I followed the next week-end and we both felt so much better for a couple of nights' undisturbed sleep, that we told ourselves we would do this whenever possible.

At about 10.45 a.m. on November 8th, I was at work in my office when I received a telephone call from the contractor on my West Division. He said he had just returned from the division and was speaking to me from the village. He stated that my young European assistant had been fired on. I made immediate contact with the police who arrived very promptly together with a detachment of Scots Guards. With them I went to the scene of the alleged shooting. It was all too true. There on a hairpin bend on the estate road was the burnt-out wreckage of the motor cycle which my assistant had been riding. We found 25 rounds of expended Sten gun ammunition and two of .45. After a prolonged search we found his dead body in a near-by swamp.

Next day, November 9th, as I was leaving Cheras Road cemetery after the burial of my assistant, I was informed that

Tapping rubber, *c.*1910. Nowadays only one 'panel' of the tree-trunk, not four, would be tapped at any one time, from Ashley Gibson, *The Malay Peninsula and Archipelago*, J. M. Dent & Sons, Ltd., London, 1928.

my small estate lorry had again been ambushed....

And so life went on and one day merged into the next. Every time I moved I had my automatic strapped round my waist and my two bodyguards accompanied me. On an evening late in November I was just about to go to bed when the telephone rang. The factory guards reported that they had seen lights moving in the rubber and that they had opened fire. The lights had then disappeared. I commended their action, advised them to remain alert and said we would examine the area at first light. In due course we did and there were the footprints of our nocturnal visitors....

I had had a particularly tiring day on January 9th and retired after an early dinner. My hope of an undisturbed night was, however, doomed to disappointment. SC Mohd Totin suddenly appeared at my bedside and all he could say was 'Api banya besar, Tuan', (A very big fire, sir). One glance at my alarm clock showed that the time was 2.30 a.m. My next glance assured me that there was indeed a big fire in the village area. To reach the telephone was a matter of seconds. The police informed me in reply that the Post Office in the village was ablaze. Next morning I visited the scene and there was the charred framework of what had been our Post Office. Standing alone and intact in the centre was the steel safe in its concrete emplacement. The Indian postmaster, his wife and six children escaped from the blazing building unhurt. Next day this official had opened a temporary office in the station buildings and 'business as usual' was the order.

Anon, 'A Planter's Life Today', *British Malaya* (monthly magazine), April 1949, p. 219.

61
On a Rubber Estate—Just Another Day

RAVINDRA JAIN

This passage is from a book written by an anthropologist who made a methodical study of the community of Indian labourers on a rubber estate ('Pal Melayu') in Selangor. For six months the author and his wife occupied a labourer's *quarters alongside the similar dwelling allotted to the Indian foreman (*kangani*), Ranganathan, and his family, whose early morning routine is here described. The household includes Ranganathan's married son (Munisami) and daughter-in-law (Kamala) and their sons, Murugan (10), Tanapal (8), Bappa (5), and Moghan (3) and a daughter, Papati (7), and also Ranganathan's unmarried son (Kesavan—aged 23).*

Although the standard of housing and amenities provided by the employer (in compliance with government regulations) had risen over the half century of the estate's existence, the general description of domestic life among Indian estate labourers would probably have fitted a labour force of a generation or more before 1963.

The total Indian population of Malaya increased from about 100,000 in 1901 to 800,000 in 1957, though it only gradually stabilized as a settled community. The majority were labourers on rubber estates.

EXACTLY at 4 A.M. on a cool starry morning in mid-1963, the siren in the factory at Pal Melayu sounded four times. Following the time signal, loud yells by the estate watchman could be heard issuing from a distance of about a hundred yards. His shouts grew progressively louder as he moved closer to the lines, flashing his torch to find his way in the dark. No one in Ranganathan's house woke up, although kerosine lamps began to be lit in house blocks here and there. An hour passed during which fires were made in some of the houses. The noise made by the washing of buckets or chopping of wood mingled with the cackling of geese, crowing of cocks, and occasional cries of children. A few women could be seen strenuously grinding chilies and turmeric on the stone mortars to prepare paste for the day's curry. One woman poured tea into a thermos flask to carry to the tapping task.

At 5 A.M. another time signal sounded from the factory. Ranganathan was the first to wake up in his house. He began shouting 'Murugan', the name of his eldest grandson. Covering his head with the kangany's headcloth, he set out to wake up members of his gang, shouting high-pitched abuses at the top of his voice. Almost simultaneously, a large number of neighbors woke up. Kamala came downstairs and lit the fire on the hearth where on the preceding night Kesavan had left firewood in readiness for the morning. From water stored in a large drum of corrugated iron, Kamala filled an aluminium kettle and put it on the fire to boil water for tea. Three-year-old Moghan, still half-asleep in the room upstairs, kept whining all the time. As Kamala went upstairs to bring condensed milk, teadust, and sugar (all household provisions except condiments are kept in a cupboard upstairs), she attended briefly to Moghan. Before coming downstairs she called out to Murugan that tea would be ready. This was meant as an indication for all members of the household to come down.

Munisami came down next carrying Moghan in his arms. Kamala began winnowing rice. One by one all children and adults came down and cleaned their teeth with a pinch of charcoal powder kept in a paper bag. Kamala instructed her eldest son to clean the empty beer bottle in which she carried her tea to the task. All those who must go to work that morning—Ranganathan, Munisami, Kesavan and Kamala—drank their tea in enamel mugs, standing or squatting near the fire. Kamala shared her tea with Moghan. She next filled the clean bottle with tea. After the elders had finished, Murugan made tea for himself and his younger siblings. Papati [his sister] was asked to clean the mugs. Kamala filled an aluminium pot with rice and water and put it on the fire to cook. She instructed Muniammal, the twelve-year-old non-school daughter of neighbors, to watch the rice and remove it from the hearth when it was cooked.

It was now about 5.45 A.M. and from a transistor radio, turned up full pitch in a nearby line, the sound of a song from a Tamil film poured out. Most of the workers had already left for the muster ground, and the working members of Ranganathan's household were ready to leave. Munisami, in his tapping shirt and blue shorts, carried two small mat baskets for collecting latex scraps. Kesavan wore only a singlet over a pair of shorts. Each brother had a bicycle and two buckets. One bucket was tied to the carrier at the back and the other hung on the handle. In one of the buckets the brothers had put their tapping knives. Kamala dangled her two buckets in hooks attached to the two ends of an arched bamboo stick. She placed the bamboo across her left shoulder and left for the muster ground on foot. Ranganathan with his inevitable head-cloth and a shoulder bag containing the kangany's 'tools', also left for the ground on foot. By 6 A.M. the household was in charge of the children.

A few minutes before 7.30 A.M., Murugan, Tanapal, and Papati left for school, leaving Bappa and Moghan behind. Both these children could have been left in the nursery, but Bappa was now regarded as sufficiently grown-up to take care of himself and of his younger brother as well. In fact, of late Kamala's efforts to leave both of them, or only Moghan, in the nursery had proved futile. As I gathered from the nursery caretaker, my other neighbor's wife, Bappa was no longer amenable to her discipline and Moghan felt miserable staying in the nursery without his elder brother. On the other hand, throughout the period they were left alone, the brothers played together or in the company of other children of their age group, similarly left by their parents. The neighbor's daughter Muniammal kept an eye on the children.

Ravindra K. Jain, *South Indians on the Plantation Frontier in Malaya*, Yale University Press, New Haven, 1970, pp. 63–6,

People

62
Raffles at Malacca

MUNSHI ABDULLAH

Munshi Abdullah was born at Malacca in 1796. He was a boy of fifteen when, in 1811, the British forces which were to occupy Java assembled at Malacca. Stamford Raffles, destined to be Lieutenant Governor of Java (1811–16), came from Penang to Malacca as civilian political adviser to the Governor-General of India, Lord Minto, in planning the operation. Abdullah, whose autobiography is one of the Malay classics, here describes his first encounter with Raffles.

SOMETIME later a rumour was heard in Malacca that the English were going to attack Java. Two or three months after we first heard this news Mr Raffles and his wife suddenly came to Malacca, with an English copying clerk named Mr Merlin and a Malay clerk named Ibrahim, a half-Indian from Penang. Mr Raffles took a house in Bandar Hilir on an estate owned by the Chinese Kapitan's son whose name was Baba Cheng Lan. He brought with him many rare objects of European workmanship, things displayed in cabinets, pistols, costly satin materials and gold-embroidered muslin, and a great many things I had never seen before; many broad-cloths of great fineness, ornate clocks, and papers for writing letters to Malay rulers and princes with gold and silver head-ings, and many other objects intended as presents for Malay royalty.

One day Ibrahim the Malay writer came to my house and sat

talking about how Mr Raffles was looking for copyists whose hand-writing was good, and how he wanted to buy old Malay letters and texts. He said that those who had any should take them to Mr Raffles's house at Bandar Hilir. One of my uncles named Ismail Lebai had very good hand-writing, and he and his younger brother Mohammed were both taken on as copyists. The next day Ibrahim came again and asked for a specimen of my hand-writing. After I had written one he took it to Mr Raffles, and the same afternoon one of his attendants came to summon me. So I went along, and Mr Raffles said to me: 'Copy these letters into a book.' Now working there was a Malacca-born friend of mine named Tambi Ahmad bin Nina Merikan. There was all manner of work being done; some copied stories, some wrote letters, others wrote about the idioms of the Malay language, its poetry and so on. Each of us had his own task.

Now as to Mr Raffles's physical features I noticed that he was of medium build, neither tall nor short, neither fat nor thin. He was broad of brow, a sign of his care and thoroughness; round-headed with a projecting forehead, showing his intelligence. He had light brown hair, indicative of bravery; large ears, the mark of a ready listener. He had thick eyebrows, his left eye watered slightly from a cast; his nose was straight and his cheeks slightly hollow. His lips were thin, denoting his skill in speech, his tongue gentle and his mouth wide; his neck tapering; his complexion not very clear; his chest was full and his waist slender. He walked with a slight stoop.

As to his character, I noticed that he always looked thoughtful. He was very good at paying due respect to people in a friendly manner. He treated everyone with proper deference, giving to each his proper title when he spoke. Moreover, he was extremely tactful in ending a difficult conversation. He was solicitous of the feelings of others, and open-handed with the poor. He spoke in smiles. He took the most active interest in historical research. Whatever he found to do he adopted no half-measures, but saw it through to the finish. When he had no work to do other than reading and writing he liked to retire to a quiet place. When he was occupied in studies or conversation he was unwilling to meet anyone who came to the house until he had finished. I saw that he kept rigidly to his time-table of work, not mixing one thing with another. I noticed also a habit of his in the evening after he had taken tea with his friends. There was an inkstand and a place for pen and paper on his large writing-table, and two lighted candles. After he had walked to and fro for long enough he would lie on the table on his back staring upwards, and close his eyes as though asleep. Two or three times I thought he was actually asleep, but a moment later he would jump up quickly and start writing. Then he would again lie down. This was his behaviour every night up to eleven or twelve o'clock when he went to bed. Every day it was the same, except occasionally when his friends came in. When morning came he would rise and fetch what he had been writing the night before, and walk up and down reading it. Out of ten pages he would take perhaps three or four and give them to a writer to copy out. The rest he would tear up and throw away.

He employed four men to search for specimens of natural history. One he told to go into the jungle and look for various kinds of leaves, flowers, fungi, mosses, and so on. Another he told to find worms, grasshoppers, various kinds of butterflies, beetles, and other different insects, cicadas, centipedes, scorpions, and the like, and he gave him some needles and told him to set the specimens. Another man he despatched with a basket to get coral, various sorts of shells, molluscs, oysters and the like, and also fish. The fourth man went out catching wild animals like birds, jungle fowl, deer, and small quadrupeds. Mr Raffles kept a large book having very thick pages in which he used to press leaves and flowers and the like. Anything which could not be inserted between the pages he gave to a certain Chinese from Macao who was very expert at drawing life-like pictures of fruits and flowers, telling him to copy them. He also had a large barrel full of some sort of spirit, possibly toddy or brandy, in which he put live animals such as snakes, centipedes, scorpions and the like. Two days later he would take them out and place them in bottles, where they looked just as if they were alive. People in Malacca were surprised to see such a thing, and many were able to earn good money searching for the creatures of the sky, the land and the sea; of the uplands, the lowlands and the forest; things which fly or crawl; things which grow and germinate in the soil; all these could be turned into ready cash. There were also people who brought Malay manuscripts and books, I do not remember how many hundreds of these texts there were. Almost, it seemed, the whole of Malay literature of the ages, the property of our forefathers, was sold and taken away from all over the country. Because these things had money value they were sold and it did not occur to people at the time that this might be unwise, leaving them not a single book to read in their own language. This would not have mattered if the books had been printed, but these were all written in longhand and now copies of them are no longer available. There were some three hundred and sixty books in all, apart from *Shaer* and *Pantun* and other kinds of verses. Yet other books Mr Raffles borrowed and had copied, keeping four or five copyists employed on this task alone.

Abdullah bin Abdul Kadir, Munshi, 'The Hikayat Abdullah', translated by A. H. Hill, *Journal of the Malayan Branch of the Royal Asiatic Society*, Vol. XXVIII, Pt. 3, 1955, pp. 72–4.

63
The Imam of Pulau Aur

JOHN THOMSON

John Turnbull Thomson was a surveyor who entered the government service in Singapore in 1841. His duties included a survey of the eastern approaches to Singapore, in connection with the project for erecting the Horsburgh Lighthouse. In the course of his work he visited the islands off the east coast of Johore, including Pulau Aur. After his retirement, Thomson wrote two entertaining books on life in the Straits Settlements, entitled Some Glimpses into Life in the Far East *and* Sequel to Some Glimpses into Life in the Far East *in 1864–5. But this passage is from an article on Pulau Aur which he contributed to the* Journal of the Indian Archipelago *in 1850.*

The satirical picture of this Malay reveals as much of the European attitude as of the man himself. But there was a class of Malay notables and traders whose travels in South-East Asia had widened their knowledge, though without perhaps improving their character.

WE first proceeded to the house of the hereditary Imam, or high priest of the island, where there were great crowds of natives waiting to receive us. We found him sitting on an old rusty 24 pounder, several of which were lying on the beach pointed seaward and said to have belonged to a brig that was cast away on the reef 30 or 40 years ago. He conducted us to his palace which is built on posts and covered with leaves, carpets were spread for us in the principal room apparently appropriated to the male visitors, and from which the apartment of the women was divided by a screen or an apology for one which, when sitting, admitted of view and communication from room to room, through a horizontal opening guarded by bars....

We found the small pox raging at the time of our visit, so considered it unsafe to enter the various kampongs. The Imam, who professed great solicitude and anxiety about this matter, begged a passage to Singapore in order to procure the vaccine virus which he said he understood how to apply....

He described himself as the High Priest of the island, the holder of all ecclesiastical authority: he possessed several papers written in English, and amongst the rest a number of the *Free Press* of Singapore, of which he was evidently very proud; in these he is styled the Prince of Pulo Aur, uncle of the Rajah etc. Amongst the scraps are a rude portrait of himself, and a certificate of his good disposition towards the English; above all these he places a rude advertisement that he sells the best wood and water to be procured at Pulo Aur and solicits the custom of Captains visiting his jurisdiction. It will be correctly surmised that his title of head of the church would convey rather an erroneous idea of the subject to the western reader....

His small native island has not afforded scope for his versatile tastes, and he has traded from one end of the vast East Indian Archipelago to the other—from the Papuas of New Guinea to the Battas of Sumatra. He has also acted the courtier at the palaces of Pontianak, Pahang and Lingga. As a warrior he has fought against the King of Palembang and the Rawas of the Malay Peninsula. In his conjugal ties he has had varied experience. He has stood at the hymeneal altar seven times. As his first love he claimed a daughter of his native island, but forsook her for a fair one at Pahang. He next espoused a lady attached to the court of Pontianak. Her he deserted for a belle of Siantan in the Anambas group, the mother of his only child. Next he transferred his affections to a fair one of Sidili. Another he courted and won at Palembang, and lastly he married a slave of Wan Syed of Pahang, but when we found him he was living with his spouse of his younger days at Pulo Aur. To visit all his wives a journey of several thousand miles would have to be undertaken. With all this he was not backward in narrating his experiences amongst the nymphs in the purlieus [outskirts] of Singapore....

On the first day of our taking the Imam on board, and before our close quarters had made us better acquainted we anchored in the Bay of Joara on the east shore of Tuman [Tioman Island]. As we were dropping anchor the Imam accosted me, 'Sir, as I am the chief of this kampong and as I am also high priest, it will be necessary for you to fire a gun and send your boat to acquaint the people that the Tuan Imam Ahmat Bin Abdullah of Pulo Aur is here who desires them to wait on him.' This was refused but he was told he was at liberty to accompany the commander who was going on shore. He remonstrated, stating that such a proceeding was not becoming his dignity, that these people were slaves and menials of his, and that it was not consistent with his position to visit them. Seeing he could not gain his purpose he said, 'Well, at all events I shall go and prove to you in what high estimation I am held here, they are all followers of mine and as I have taken no followers from Pulo Aur perhaps they may wish to escort me to Singapore. You will I hope at all events allow the whole population to come and visit me on board.' With this he proceeded on shore, but the poor Imam came back alone much down hearted, having only been able to wring out of his beloved subjects two very small and lean chickens. These he carried in his hands and remarked they would do very well as a present to his English friends in Singapore.

He was fond of repeating Arabic texts from the Koran (a very common cloak to ignorance in other parts of the world) which he did not understand; his calendar of prophets included Adam, Moses, Christ and Mahomed. He was very superstitious and inwardly held the Hantus and Kramats,* of his native island in greater fear and reverence than the dogmas of Mahomedanism which he professed to believe. Thus one day, on his describing the forms to be observed in approaching their famous Kramat, I was led to ask him why seeing he believed in Tuan Allah he should also trust in peris, dewas and mumbangs,* for is not Allah greater than these? To this he replied, 'It is very wrong certainly but we must respect the customs of the men of old who had converse with these demigods, a thing that is denied to us ignorant children of the present age.'

As time drew on our Imam settled down to his wonted habits. The finery in which he arrayed himself on first acquaintance he threw off piecemeal, his garments at last counted two in number. A pair of light canvas trowsers not too carefully buttoned in front gave shelter to the nether man, while an old flannel jacket borrowed from a Singapore acquaintance decorated the upper; bare-feeted and bald pated he crept about the deck. In this habit I found him busy one day plucking the fowl, in which, with the assistance of goggle spectacles, he was occupied with great intensity. 'Well Imam what is this you are doing?' 'Oh, only cleaning a fowl which that satan of a Kling is too lazy to pluck.' 'But this is not proper employment for a man of your dignity.' 'I would condescend to any thing for my white friend,—besides what are we but Moniet and Kra (apes and monkeys) such employment becomes us.'

As our voyage drew to a close his humility and attention increased. One day he appeared to be more intently desirous of pleasing than I had before observed. At last he edged nearer and remarked: '... I am a poor man and would like to have some money to buy a little opium.' 'Is not the consuming of intoxicating drugs forbidden by the religion whose minister you profess to be?' 'That is true, but where is a pennyless man's conscience when the belly is empty?' 'Have you not been well fed for twenty days gratuitously, and do you mean

*Malay demonology of this period included ghosts (*hantu*), sacred objects and places (*keramat*), and spirits of various kinds (*peri*, *dewa*, and *mambang*). See Passage 70 for some examples.

to say that you have no money in your box below?' 'Satu duit tidak, not one piece. (He had 60 Spanish dollars.) I trusted entirely to the usual generosity of the Orang Puteh (white man) so came away penniless. I will be glad to take 5 dollars or 20 dollars, whatever your honor's kasihan (kindness) may dictate.' 'So the procuring of vaccine virus to allay the misery of your suffering countrymen was not your object in coming to Singapore?' 'Ha, ha, ha,—who would have believed that? God takes whom he likes and what would I gain for any trouble by a voyage for such a purpose.'

John Turnbull Thomson, 'Pulo Aur', *Journal of the Indian Archipelago*, Vol. 4, 1850, pp. 191–6.

64
Tunku Panglima Raja of Selangor

EMILY INNES

After moving to her new house on Jugra hill (Passage 44 above) Emily Innes was a neighbour of the Sultan and his circle, who had made a similar move. The Sultan's brother-in-law and confidant is here described. Tunku Panglima Raja was indeed a more or less retired pirate whose traditional views got him into the bad books of the Resident (Bloomfield Douglas) with untoward consequences for them both. The rather unlikely friendship between the old chief and the lonely housewife tells one a good deal about them both.

I sometimes had the more civilized of the rajas in to afternoon tea if they happened to call at the right time for it, and was much amused with their ways—especially with one, the Tunku Panglima Raja. He was a very fine-looking old fellow, with large, bright, piercing eyes, a high forehead, and a good aquiline nose—not the flat, wide, fleshy snub usual with Malays. He was, in fact, not of pure Malay, but of Bugis extraction, I believe, as was also the Sultan. Tunku Panglima Raja wore a black silk handkerchief on his head, stiffened with rice-starch, and twisted into a tremendous erection, something like a bishop's mitre, but with the two ends sticking up like little horns on either side. The rest of his dress consisted of a jacket, buttoned only at the neck, and showing his brown skin from thence to the waist, and a sarong, the twisted part of which was stuck full of krises, that gave him a warlike appearance.

His favourite beverage, I found, was not tea, but Bass's pale ale, which, however, he took in homoeopathic doses. He would never have more than half a wineglass of it in one day; but declared that that quantity did him a great deal of good as a pick-me-up when tired. He even joined us sometimes at luncheon, and apologized gracefully for eating with his fingers, saying he was an old man and could not learn new ways. He and I once had a little talk about the respective merits of fingers and forks, in which I confess he had rather the best of the argument.

'To tell the truth,' said he, 'we Malays do not care to eat with forks and spoons because we think it such a dirty practice. We say to ouselves, "What do I know of the history of this fork? it has been in a hundred, perhaps a thousand mouths; perhaps even in the mouth of my worst enemy." This thought is very repulsive to us.'

'But, Tunku,' said I, 'the fork is thoroughly well cleaned, or ought to be, every time it is used, first with soap and hot water, then with plate-powder.'

'*Ought to be*; quite so,' said the Tunku. 'But how do you know that your servant does not shirk his work? If you have a lazy servant, you are liable to eat with a fork that has not been thoroughly cleaned. Whereas I know that my fingers are clean, for I wash them myself before eating. They are quite as clean as the cleanest fork, and they have two great advantages over it—one, that they have never been in anyone's mouth but my own; and another, that they are never lost, or mislaid, or stolen! They are always at hand when one wants them.'

I was quite sorry when this poor raja got into disgrace with the Government for wrecking a ship that came too near his little bit of coast, and helping himself to its cargo. We had become on very friendly terms with him, and used often to receive little notes from him if he wanted any trifle, the penmanship being done not by himself, but his clerk....

Our lawn-tennis ground had now become covered with grass, and we often played for an hour *tete-a-tete*. It was surrounded by a rude fence 'to keep off the ruder natives', as one of our visitors remarked. The natives were much interested in looking on at the game, and the Tunku Panglima Raja one day expressed a wish to join us. He was very active, and flew about the ground with his petticoats tucked up, but had not much notion of the game, hitting the ball straight up into the air as high as he possibly could being his idea of playing.

Emily Innes, *The Chersonese with the Gilding Off*, 2 vols., Richard Bentley and Sons, London, 1885, Vol. 1, pp. 101–3 and Vol. 2, p. 6.

65
Sultan Yusuf of Perak in Penang

HUGH CLIFFORD

Raja Yusuf had grown up in Perak during the turbulent period of the mid-nineteenth century. He was an able man and a doughty fighter but various flaws of character and manners added to his unpopularity with his own people. When Perak came under British control in 1874, it set in train a sequence of events by which all the more likely candidates for the Sultanate were eliminated. For lack of anyone better qualified, Hugh Low (see Passage 36) persuaded a reluctant Perak to accept Yusuf as Regent. Somewhat unexpectedly, Low's diplomacy and Yusuf's realism made it a successful arrangement, leading to Yusuf's elevation to the Sultanate in the last year (1887) of his life.

It was in this period that young Hugh Clifford, a youth of eighteen, came to Perak to be trained by Hugh Low (and by Frank Swettenham who deputized for Low on occasion). Clifford here describes the agonies of escorting Yusuf into the polite society of Penang. The opening paragraph refers to the custom by which a Sultan, after his death, was referred to, not by name, but by a conventional and usually complimentary descriptive phrase, chosen with reference to his reign or personality.

HE was a typical son of the old *régime*, a barbarous person of unspeakable manners and morals. When, some years later, his time came to die, and when, in accordance with the custom of the land, his people conferred a posthumous title upon him, they called him 'Al-Merhum Rahmat Allah', which, being interpreted, is, 'The late king, *God be merciful to him!*' They felt that no conventional phrase of laudation or glorification would fit him, and that, in view of his manifold iniquities, the best that his most sanguine friends could hope was that Allah, the Merciful, the Compassionate, might grant him the forgiveness which he had not earned but sorely needed.

In the company of this potentate I spent three lurid weeks in 1884 while he disported himself in the neighbouring colony of Penang....

One of my chief troubles lay in the inability of my king to appreciate the advantages of punctuality. In common with all Malays of his generation, he held time to be valueless, and regarded an hour or two either way as a thing of no account. I remember my distress when all my efforts failed to drag him from his sleeping-mat in due time for a parade of a European regiment which, to the extreme discomfort of a peppery old commanding officer, had been ordered in his honour. British troops in Asia are very precious things, and *the* unpardonable sin is to keep them standing in the sun-glare. When, therefore, half an hour late I at last sneaked on to the ground in the wake of my king, I longed, if ever a man did, that the earth would gape and swallow me. The *raja* was a fine billow of a man, and he waddled with the rolling gait of a sailor. His figure was portly, and he had a strong predilection for gorgeous colours and barbaric raiment. On that particular morning he was chastely clad in a bright yellow cap with a scroll from the Kuran embroidered upon it in black letters; in a pink cloth coat; a pair of green silk Chinese trousers reaching to the middle of his shins; and in a number of red, blue, and purple shawls huddled about his waist. Lengths of brown hairy legs protruded from the bottoms of his trousers, and his bare splay feet were thrust into canvas tennis-shoes without strings. He leaned heavily on a long patriarchal staff, and scowled furiously at all the world, as was his engaging habit. At intervals he emitted a swine-like grunt or snort....

Sultan Yusuf, with two of his sons (seated) and two retainers, from Major Fred McNair, *Perak and the Malays: 'Sarong and Keris'*, Tinsley Brothers, London, 1878.

When invited to five-o'clock tea by some lady of high standing in the community, he would take complete charge of the tea-table, shouting to his ragamuffin followers to join in the plunder, and distributing all the available comestibles among them before I could intervene. All this he would do with a wicked eye cocked in my direction to note how I bore up under the ordeal; or else with the same iniquitous leer he would ostentatiously remove the soaking quid of tobacco, red with betel-juice, from between his lip and gums, and would cast it upon the carpet in our midst with a soft, splashy *flop* that made the stoutest shudder....

In the end I used to get him back to his own country, everlastingly disgraced, it is true, but more or less unharmed. Yet it is curious how understanding and sympathy bind a man to even the least attractive personality, for I grew to have more than a sneaking affection for my wicked old king. I learned from him much concerning the management of his people which has since stood me in good stead; I was often forced to admire the hard-bit, strong-willed, shameless, but fearless old curmudgeon; and when at length he died in the odour of iniquity, I joined heartily, and more than a little sadly, in his people's prayer, 'God be merciful to him!'

Hugh Clifford, *Bushwhacking and Other Asiatic Tales and Memories*, Harper, New York, 1929, pp. 171–7.

66
Ungku Sayid of Paloh

HUGH CLIFFORD

For much of his early Malayan career Hugh Clifford (see Passage 65 above) served in Pahang in the difficult period, from 1887, in which the Sultan and his chiefs were induced to accept external control through a Resident (a position to which Clifford rose in the 1890s). There was, however, a major Malay revolt; it was defeated but the leaders were pursued into Trengganu and then Kelantan by a party led by Clifford. In Trengganu the exiled Pahang leaders had the sympathy and blessing of the leading Islamic teacher, Ungku Sayid of Paloh, who was a man of influence and an uncle of the Sultan.

Mindful of the power of Islamic leadership in a 'holy war' such as the rising in the Sudan, and the death of General Charles Gordon at Khartoum ten years before, Clifford and other officials were apprehensive lest a similar movement, led by the Sayid, might confront them in Trengganu in 1895. But as Clifford came to realize, Islam in South East-Asia is generally less prone to militant fanaticism than in the Middle East.

THE saint lives secluded in the retirement of a shady sleep-steeped village. He is rich in flocks and herds, loves his fruit-groves and his flowers, is surrounded by a number of youths who sit at his feet and run quickly to do his bidding, and weekly he preaches after the Friday congregational prayers to throngs of devotees. The sainthood has been passed down from father to son almost since the beginning of things. A far from usual knowledge of the Muhammadan Scriptures, a gift of ready speech, a vast display of ostentatious piety, a certain asceticism of mien, are the saint's stock-in-trade. The imagination of the people, aided by the saint's own genuine and unbounded faith in his claims to sanctity, accomplish the rest. Stories are told of the little useless miracles which he has worked,—such as the introduction of tiny fish into the heart of a cocoanut, where no fish should be—are repeated solemnly, and are accepted without inquiry or proof. Great strength of character, enormous belief in himself, the long years during which the habit has been formed of dominating all men by the force of his will, have given to this man a personality impressive, powerful, magnetic. A little, shrivelled, glassy-headed man, from out whose deep sunken eyes there glares the soul of a fanatic—in modern Europe we had labelled him a 'crank'; in modern Asia, his fellows, taking him at his own estimate, know him for a saint. Therefore, when he lifts up his voice and preaches a *jahad*, bidding men battle for the Faith, the outlaws have no difficulty in gathering a respectable number of adherents so soon as they unfurl the green standard of Muhammadan war. All the best ruffians in this outlying flange of Southern Asia flock to their side; the young bloods who are always 'spoiling for a fight', the sweepings of the Benighted Lands, the men with nothing to lose and everything to gain, join them. The calls of religious enthusiasm might, perhaps, be withstood; but when coupled with a certainty of victory—for has not the saint foretold the event?—and the prospect of unlimited loot, they are found to be irresistible. The saint distributes charms against knife-thrust and bullet-wound, scratches texts from the Kuran on the blades of weapons with a rusty nail, and predicts the triumph of the forces of the Faith. Then he retires once more to his fruit-groves, his flowers and his devotees, to fast rigorously and pray for success and for a great slaughter of the infidels.

Hugh Clifford, *Bushwhacking and Other Asiatic Tales and Memories*, Harper, New York, 1929, pp. 72–4.

67
Frank Swettenham and Albert Braddon

ARTHUR KEYSER

Eccentricity flourished among Europeans in the tropics. It was partly a form of self-indulgence, since if a man chose to be odd he was often in a position to do as he pleased.

Frank Swettenham, one of the outstanding Malayan administrators of the late nineteenth and early twentieth century, was a careerist who kept his affectations within bounds. Keyser began his Malayan service with a few months in Kuala Lumpur in 1889, where Swettenham was British Resident. He mentions that during the 1914–18 war, Swettenham, long since retired from Malaya, worked immensely hard (in his sixties) as Joint Director of the Official Press Bureau, which among other functions censored press telegrams.

From Selangor Keyser moved on to Jelebu (see Passage 11 above). Here he met Dr Albert Braddon, then State Medical and Health Officer, Negri Sembilan. Braddon did not discover 'the causes and cure of beri-beri', a scourge which caused many deaths among the immigrant Chinese labourers. But he did make the decisive breakthrough by establishing that beri-beri arose from dietetic factors connected with rice. Unlike Swettenham, Braddon indulged his foibles to the point at which cantankerousness denied him the success which his talents might have brought him. He left the government service, made and lost money in tin-mining and died in comparative poverty.

Frank Swettenham

IT was a weary, hot voyage up-stream, a route long since covered by a railway, until finally Kuala Lumpor, the capital, was reached. And there, standing on the bank was a picturesque figure, surmounted by the enormous hat which he then affected—Frank Athelstane Swettenham, one of England's Empire Builders.

After a kindly welcome we drove to the Residency. A few hours later my host in that delightful drawl, which always made me wonder whether its words were really as soft as their sound, casually enquired whether I had come out to work or play, since, when once he knew, he could arrange a programme for either....

I commenced work as Private Secretary to my Chief. In this position it was once my duty to interview a young applicant for a place. Obtaining from him a full account of his qualifications, all most excellent, I went beaming to repeat them to the Resident. When I paused, without looking up from his writing, he asked, 'Can he play cricket?' I begged leave to retire and find out. As the answer obtained was satisfactory, the candidate was engaged, and he was started on our little groove. I naturally watched his first innings at cricket with some nervousness....

Sir Frank Swettenham used to affirm that acre for acre, above ground and below, the Malay Peninsula was a hundred per cent richer than any other country in the world....

I can visualise that bend in the narrow path from Jerang to Jelebu, where (we were necessarily riding in single file) my chief, above-named, looked over his shoulder and casually remarked that he thought my report on the Keeling Cocos Islands one of the best of the kind which even he had ever read. As I write, I see distinctly the gigantic tree fern, overhanging dismal swamp, under whose shade my hat seemed suddenly to be too small for my head....

When news came of his K.C.M.G., and as I congratulated him, supposing that he would now be Sir Francis, he replied that the supposition was surely a strange one, seeing that his Christian name was Frank.... Long years after, during the war, when visiting this distinguished official, then Press Censor, in his room at Whitehall, I asked him for a post; to which he replied that perhaps he could suit me, but was I prepared to work all night as well as all day? ... Again, when a few months ago I lunched with my old friend in Seymour Street, how pleasant it was to see him surrounded by treasures gathered in his life abroad, and note the comfort and atmosphere of his home, which proved that the tranquillity well earned was now at his call!

Dr Albert Braddon

Dr Braddon, an officer of unusual ability and originality (well known amongst the discoverers of the causes and cure of Beri-Beri) ... recognised no fetters of convention, not even those marked by the clock, since when it was dark it was sufficient to light a lamp. A sun helmet in the daytime he found too hot, but needed it at night to keep off the dew.

He would probably diagnose your complaint as 'ailment', which the native hospital dresser might cure, adding that in 'illness' he would be willing to attend himself. As to most complaints of the body, however, he believed that no one could know more than the person who owned it.

He disliked 'kickshaws' [elaborate cookery], saying that a piece of meat should be hung to the ceiling by a string so that each diner could cut chunks from it with a knife. And on one occasion, after dining off the efforts of my excellent Chinese cook, mostly minced, he complained that he found his teeth useless, and the next time I asked him to dinner he would bring a funnel.

But he appreciated culinary skill, and in his own house had been known to call his cook at the conclusion of dinner and order him to give him another and better one. This he would wait for and eat. And as there was no record of time, this original officer would rise at any hour of the night and ask for a meal.

Dr Braddon so scorned red tape that he was described as a buccaneer official by his Chief, but he was also so clever that few ventured to engage him on paper. He was a fearless rider, a very necessary accomplishment at that time, as we imported and rode unbroken colts from Australia....

Memories of Dr Braddon will always be associated with a day when I was with him, at Kuala Pilah, capital of a neighbouring State, and the local authorities, taking advantage of the presence of a medical man, asked us to remain and witness the execution of a Malay, sentenced to death for the murder of a Chinaman. The culprit was led forth, and the executioner, passing the curved blade of his kris gently through his fingers, addressed him and said: 'I wish you to take notice that it is not I who am going to kill you, but God, whose humble instrument I am. You murdered the Chinaman and must pay the penalty.' To which the Malay convict, quite unmoved, replied: 'I did kill the Chinaman, you intend to kill me, please do so and refrain from boring me with your chatter first.' Then the executioner plunged the point of the weapon into the left shoulder downwards to his lungs, and death ensued at once.

During our long ride home this unpleasant scene was discussed with my companion. I described it as barbarous, while Braddon called it humane. Towards the end of our journey, however, he suggested changing sides, saying that, of course, I was right, the killing had been perfectly horrible, and he would now make out a stronger case against it than I had been able to do.

A. L. Keyser, *People and Places: A Life in Five Continents*, John Murray, London, 1922, pp. 100–1, 133–5, and 195; A. L. Keyser, *Trifles and Travels*, John Murray, London, 1923, pp. 121–4.

68
Chinese Occupations

DANIEL VAUGHAN

Vaughan, like other European observers (see Passages 25 and 33 above), noted the immense variety of trades and crafts carried on by Chinese in the overcrowded streets of Malayan towns. In the latter part of this passage Vaughan draws on his experience of police and legal work (Passage 49 above) to describe the ingenuity of Chinese petty criminals.

THE Chinese are everything; they are actors, acrobats, artists, musicians, chemists and druggists, clerks, cashiers, engineers, architects, surveyors, missionaries, priests, doctors, schoolmasters, lodging house keepers, butchers, porksellers, cultivators of pepper and gambier, cakesellers, cart and hackney carriage owners, cloth hawkers, distillers of spirits, eating house keepers, fishmongers, fruitsellers, ferrymen, grass-sellers, hawkers, merchants and agents, oilsellers, opium shopkeepers, pawnbrokers, pig dealers, and poulterers. They are rice dealers, ship chandlers, shopkeepers, general dealers, spirit shopkeepers, servants, timber dealers, tobacconists, vegetable sellers, planters, market-gardeners, labourers, bakers, millers, barbers, blacksmiths, boatmen, book-binders, boot and shoemakers, brickmakers, carpenters, cabinet makers, carriage builders, cartwrights, cart and hackney carriage drivers, charcoal burners and sellers,

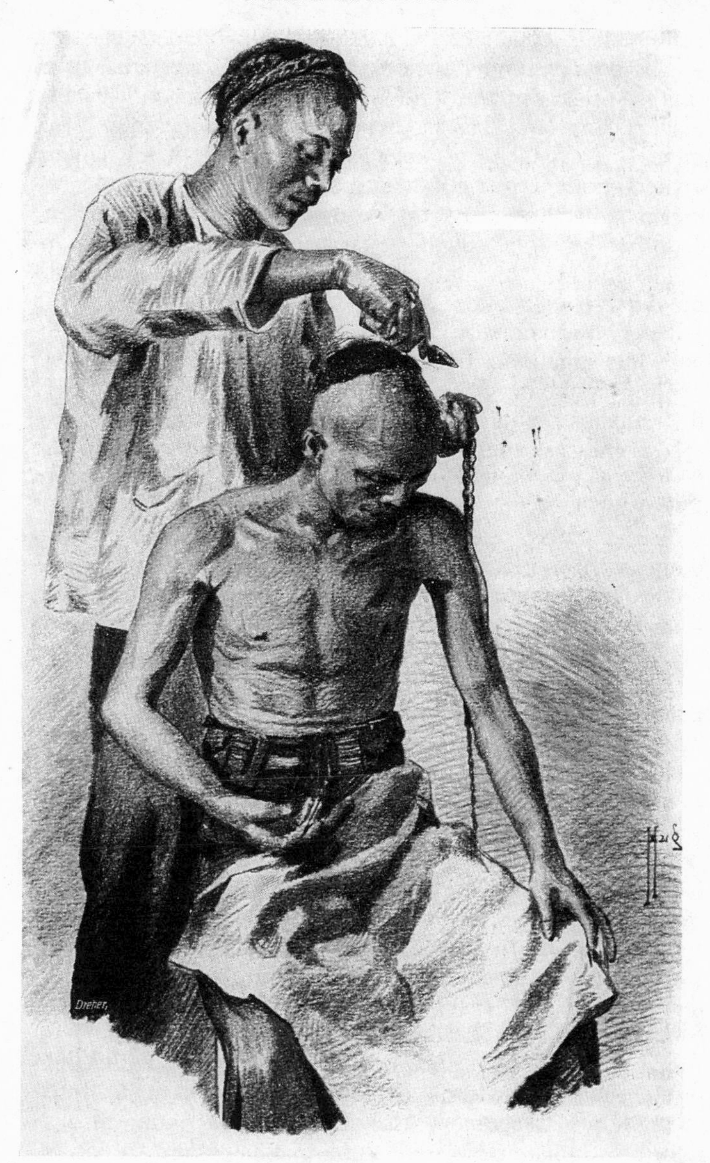

Chinese barber at work, from Hugo V. Pedersen, *Door Den Oost-Indischen Archipel: Eene Kunstreis*, H. D. Tjeenk Willink & Zoon, Haarlem, 1902.

coffinmakers, confectioners, contractors and builders, coopers, engine-drivers, and firemen, fishermen, goldsmiths, gunsmiths and locksmiths, limeburners, masons, and bricklayers, mat, kajang and basket makers, oil manufacturers, and miners. To which we may add painters, paper lantern makers, porters, pea grinders, printers, sago, sugar and gambier manufacturers, sawyers, seamen, ship and boat builders, soap boilers, stone cutters, sugar boilers, tailors, tanners, tinsmiths and braziers, umbrella makers, undertakers and tomb-builders, watchmakers, water carriers, wood cutters and sellers, wood and ivory carvers, fortune-tellers, grocers, beggars, idle vagabonds or samsengs, and thieves.

They are the most stealthy of thieves, and, it is said, stupefy their victims by burning some narcotic herb under their nostrils to increase their drowsiness, but the writer doubts the story. A petty shopkeeper in Penang was about to return to China and saved about a thousand dollars. These he packed in paper parcels each containing fifty dollars, and placed them at the bottom of his box. To make his wealth more secure, he slept nightly on the box with the key thereof tied round his waist; on a morning or two before his intended departure he found himself lying on the floor, his box open, all the dollars gone, and the key of the box taken from his person and left in the key-hole. The door of his house had been broken open and he removed from the box, the key taken from his waist, the box opened and every dollar removed without disturbing him, or his servants who slept in an adjoining room. The writer was then in charge of the police, and disbelieved the story *in toto*; it seemed utterly incredible. Fortunately, the owner had written his name and put his chop, or seal, on each parcel of dollars which subsequently led to the detection and conviction of the thieves. Two suspicious characters had visited the shop frequently on the pretence of making purchases and had sat there each time quietly obtaining information, from the unwary shopman which served them to good purpose. The dwelling of these two men was discovered and detectives set to watch them. It was ascertained that they were preparing also to return to China, and were making purchases and had a box or two packed ready for embarkation. On the day they were about to leave a descent was made on their shop, their boxes searched and nearly all the packets of dollars found with the prosecutor's mark name and chop intact; the thieves lodged opposite the principal police station.

On another occasion, thieves entered a bedroom in which the master of the house slept with his wife. To one of the legs of their fourposter was chained an iron safe containing their jewellry and money. The thieves entered by a window at the foot of the bed, filed the chain through, a pretty thick one, lowered the safe out of the window without disturbing the owner or his wife who slept soundly till the next morning, when they got up at the usual hour and found their wealth gone. The safe was found on the esplanade close to the Fort with the lid smashed in and contents gone; lying not far from an European sentry who had been on guard all night.

A wealthy Spanish lady at Penang was robbed of all her jewels worth several thousand dollars from the wardrobe which stood close to her bed where she lay sleeping in happy unconsciousness; the thief entered a window close to the bed. The writer had a thief come into his room at Singapore and carry off all his wife's jewels that she had placed on the dressing table close to the bed, besides other things without disturbing himself or wife. On another occasion a thief entered the writer's bedroom by a window within a foot of his bed, took all his clothing and hats that were hanging on a stand not far off, as well as several other things from a wardrobe; took all into the verandah selected the best hats and clothing, left the rest on the floor and decamped without disturbing any one.

Only a few months ago a lady in Singapore had a gold watch and chain taken from under her pillow whilst she and her husband lay in bed fast asleep. The Inspector General of Police also with a policeman on duty at his house was robbed of a clock and other things including a loaded revolver which hung on the tester of his bed ready for use, and neither he nor his wife were disturbed during the night; although the thief must have been a considerable time in the house and had frequently passed and repassed their bed. In many cases of house robbery there is reason to believe that the thieves are one's own servants. They sleep on the premises and often move about the house at night without exciting the suspicion of their masters, and take the opportunity some night of removing property and making it appear that thieves had entered the house.

The writer was robbed of all his plate a night or two after Christmas (1876) and in the morning it was discovered that a venetian of one of the dining room doors had been cut through to make it appear that some person had entered the house from without; but it was perfectly clear the venetian had been cut from the inside, and that it was the work of some one in the house. There was not the slightest doubt that the robbery had been committed by the servants. The property was not recovered nor was the act brought home to any of the servants.

J. D. Vaughan, *The Manners and Customs of the Chinese of the Straits Settlements*, Mission Press, Singapore, 1879, pp. 15–7.

69
Loke Yew

JOHN ROBSON

Of all the 'rags to riches' stories of nineteenth-century Chinese immigrants, the personal history of Loke Yew is the most remarkable. He became one of the wealthiest of his kind but unlike other Chinese millionaires of his time he was not shut into the traditions of the immigrant Chinese community. Although he was a shy man who avoided the limelight, he had a breadth of vision and an intellectual and commercial talent for innovation which enabled him to communicate with an alien regime.

He had a gift for making friends with, and using the talents of, other people such as Choo Kia Peng (see Passage 20) and John Robson (see Passage 19) from whose memoirs the following character study of Loke Yew is taken. The revenue 'farms' in which he made his fortune were a system by which Chinese businessmen tendered a fixed price (for three years) for the right to collect and retain government revenues from specified taxes.

THIS remarkable man was the only son of an agriculturalist in China who lived to about the age of a hundred. It was from his father that the late Loke Yew inherited his love of the soil. He came to Singapore when he was a boy, worked in a shop and saved $99 in four years. Later he went to Perak where he made and lost money in tin-mining and trading. He was one of the contractors for food supplies to the troops during the Perak war. From Perak he came to Selangor—where he spent the greater part of his life. He leased Farms (gambling, spirit and pawnbroking) from the Government, was the first man in Selangor to establish an electric power plant on a mine, constructed the Sungei Besi and Bentong roads, opened up rubber estates, started cement works and a coconut oil mill and put a lot of his money in real estate—both here and in Singapore. Charity was one of his virtues. The Hong Kong University benefitted by his generosity and it was from this university that he received his honorary L.L.D. I first got to know him well when I was Assistant District Officer at Rawang, where he had important mining interests.... In later years, after I had started a daily paper in Kuala Lumpur, I gave up journalism to become his local real estate agent—a business connection which lasted till his death in 1917.

He was married four times. The first was a child marriage in China. I was on terms of family friendship with his other wives, including the lady who survives him. Three quite different types of women, but all kind-hearted and considerate. Loke Yew had an attractive personality. I think the charm of the man lay partly in the simplicity of his nature, partly in his extraordinary ability (although he had had no education) and partly to the natural courtesy of his manner and his innate kindness of heart. I hope and believe that the only bad turn I ever did him was when, in the long ago, I sold him a horse of a very uncertain temper. Later on, when I ventured to enquire how the animal was behaving, he said, 'Oh, all right' and then as an afterthought 'last week it knocked down a lamp post.'

Careful in Small things

The towkay was always a good and kind husband, although, owing to his dislike of ostentatious display and lack of interest in entertainment or amusement, it is doubtful if he ever quite realized that the joy of life to a woman is bound up to some extent in the paraphernalia of the home, pretty things, the luxuries of modern life, and some amusement. He was generally the least expensively dressed man in his own office, kept and used old motor cars which other people would have sold, and often went backwards and forwards to his office in a rikisha. Like so many rich men, he was very careful about petty expenditure, and most generous in big things. It was the same in business. When I was his land agent we once settled an £80,000 mortgage in three minutes. The only point he wanted to discuss was the rate of interest to be charged. On another occasion when I had to advocate, say, the spending of a few hundred dollars on house repairs, the conversation might last half an hour. Perhaps his trips to Europe, his family life and the actual hardwork he did on his estates gave him more pleasure than anything else. There is a story told of the late Mrs Loke Yew (his third wife) going out in the car to fetch him home from Hawthornden estate because it was raining, and finding him wet through with a *changkol* in his hand showing a coolie how to dig. I believe that if Loke Yew had had a first class education he might have been one of the rulers of China.

Belief in God

I don't know that he professed any particular religion, but he had a very strong belief in an all-wise, all-powerful God. He once told me he would far rather have a C.M.G. from *Tuan Allah* when he died than any decoration he might get on earth. And when told about some swindle or other would say 'Well, *Tuan Allah* will judge.' He once told me he was getting 60% of the profits derived from a certain mine. Thinking he might have partners in this particular venture, I said, 'Who gets the balance?' He smiled and replied 'Ah, that is what I should like to know.' Hundreds of people went to him for financial assistance of one sort or another; many of them under the impression that it was the duty of the rich man to grant all such requests. Genuine distress seldom appealed in vain, but he always used his own discretion. In earlier life he had been a fairly heavy opium smoker, but gave it up after his first trip to Europe. Later on he became a great cigarette smoker. Neither drink nor cards ever appealed to him. He was, on balance, a successful tin miner. However it was always the land he loved. He was practically born an agriculturalist and he died an agriculturalist. If there was any one thing he disliked more than another, it was the cutting down of a coconut tree or a fruit tree. To propose the demolition of an old house was a trifle compared with any suggestion to cut down useful trees. He was always spoken of in the office as 'the old man'. He would listen to any amount of advice, but, having a very decided will of his own, it was always problematical, if he would take it. His Malay was not always easy but there was one reply well known to his personal staff and that was 'Tausa' [not worth the trouble].

Courteous and Considerate

The Towkay was undoubtedly proud of his wealth but it did not bring him unalloyed happiness. The very last time he was in my office, a few weeks before his death, he told me he was no happier with his wealth than when he had had less. I urged him, as I had done before, to give up his incessant work and worry and to be content with what he had (millions). He got up, walked about the room and said, 'Yes, I know, I know, but I can't let go' and went through a pantomime display of holding on to a big rope.

J. H. M. Robson, *Records and Recollections (1889–1934)*, Kyle Palmer, Kuala Lumpur, 1934, pp. 29–32.

70
A Dutiful Wife—and a Tall Story

WILLIAM MAXWELL

William Maxwell (1846–97) was, like Swettenham (Passage 67 above) an outstanding colonial administrator and also an eminent scholar and historian. In 1876 Maxwell led a party of Malays across Upper Perak, then almost untouched by external influences, in pursuit of fugitive Malay leaders involved in the 'Perak War' of late 1875. One of Maxwell's particular interests was traditional Malay stories and legends; he found time during his expedition to record in his journal the local legend which follows.

JAMBAI was once the abode of a celebrated family, if Perak legends have any foundation, and I affirm that if the following story seems uninteresting in its English dress, it is because the adjuncts of open air and Malay scenery are wanting.

Che Puteh Jambai and his wife were very poor people, who lived many generations ago at Pulo Kambiri on the Perak river. They had so few clothes between them that when one went out the other had to stay at home. Nothing seemed to prosper with them, so leaving Pulo Kambiri, where their poverty made them ashamed to meet their neighbours, they moved up the river to the spot since called Jambai. Shortly after they had settled here Che Puteh was troubled by a portent which has disturbed the slumbers of many great men from the time of Pharaoh downwards. He dreamed a dream. And in his dream he was warned by a supernatural visitant to slay his wife, this being, he was assured, the only means by which he could hope to better his miserable condition.

Sorely disturbed in mind, but never doubting that the proper course was to obey, Che Puteh confided to his wife the commands which he had received, and desired her to prepare for death. The unhappy lady acquiesced with that conjugal submissiveness which in Malay legends as in the 'Arabian Nights' is so characteristic of the Oriental female when landed in some terrible predicament. But she craved and obtained permission to first go down to the river and wash herself with lime juice. So taking a handful of limes she went forth, and, standing on the rock called Batu Pembunoh she proceeded to perform her ablutions after the Malay fashion. The prospect of approaching death, we may presume, unnerved her, for in dividing the limes with a knife she managed to cut her own hand and the blood dripped down on the rocks and into the river; as each drop was borne away by the current, a large jar immediately rose to the surface and floated, in defiance of all natural laws, *up-stream* to the spot whence the blood came. As each jar floated up, Che Puteh's wife tapped it with her knife and pulled it in to the edge of the rocks. On opening them she found them all full of gold. She then went in search of her husband and told him of the treasure of which she had suddenly become possessed. He spared her life, and they lived in the enjoyment of great wealth and prosperity for many years. Their old age was clouded, it is believed, by the anxiety attending the possession of a beautiful daughter, who was born to them after they became rich. She grew up in the perfection of loveliness, and all the Rajas and Chiefs of the neighbouring countries were her suitors. The multitude of rival claims so bewildered the unhappy parents that, after concealing a great part of their riches in various places, they disappeared and have never since been seen. Their property was never found by their children, though, in obedience to instructions received in dreams, they braved sea-voyages and went to seek for it in the distant lands of Kachapuri and Jamulepor.

Several places near Jambai connected with the legend of Che Puteh are still pointed out; at Bukit Bunyian the treasure was buried and still lies concealed. A deep gorge leading down to the river is the ghaut [narrow path] down which Che Puteh's vast flocks of buffaloes used to go to the river. Its size is evidence of the great number of the animals, and, therefore, of the wealth of their owner. Two deep pools, called respectively Lubuk Gong and Lubuk Sarunai, contain a golden gong and a golden flute which were sunk here by Che Puteh Jambai. The flute may sometimes be seen lying on one of the surrounding rocks, but always disappears into the depths of the pool before any mortal can approach it.

William Maxwell, 'A Journey on Foot to the Patani Frontier', *Journal of the Straits Branch of the Royal Asiatic Society*, Vol. 9, pp. 24–5.

Ceremonies and Recreations

71
The Durbar of the Malay Rulers in 1897

FRANK SWETTENHAM

Both the Malay Rulers and the colonial administrators were by tradition inclined towards public ceremonial and entertainment, which in different forms became a significant part of public life in Malaya in the first half of this century. This passage (and the next two) describe such ceremonies and celebrations. The entertainments described below were part of the first gathering (called a 'durbar') of the Rulers of Perak, Selangor, Negri Sembilan, and Pahang at Kuala Kangsar, the royal capital of Perak, in July 1897. It was deemed appropriate to assemble the Rulers of the newly formed Federated Malay States, attended by their Residents and Malay advisers, to meet the High Commissioner (Governor) and the Resident-General (Swettenham), although little serious discussion took place. There was a second durbar at Kuala Lumpur in 1903, which was rather more productive, and then a new FMS Federal Council held its inaugural meeting, again with much pomp at Kuala Kangsar in 1909.

Although Malay Rulers occasionally paid visits to each others' capitals, there was in 1897 no precedent in Malaya for bringing four Rulers together in one place. There was some apprehension (unfounded in the event) that there would be disputes over the precedence of the Rulers among themselves and so, to induce a more relaxed atmosphere, there was an elaborate programme of amusements. A number of things went wrong in the planned arrangements, but thanks to the munificent hospitality of the host Ruler, Sultan Idris of Perak, everyone had a splendid time.

THE original intention was to get all the Sultans and Chiefs to Kuala Kangsar by the 12th of July, in order that each might be separately received with due distinction, and that there might be time to settle them in their quarters before the arrival of the High Commissioner. These arrangements were upset; first by the Sultan of Pahang declining to trust himself to such a small and uncomfortable vessel as the Perak steamer *Mena*; and, secondly, owing to an accident to the steamer conveying the Negri Sembilan party. The vessel got aground in the Perak River and the consequence was that the Yang di-per-Tuan of Sri Menanti and his Chiefs did not arrive in Kuala Kangsar till the middle of the night of the 12th.

The Sultan of Selangor, accompanied by the Resident, the Raja Muda, and all the members of the Selangor State Council, duly arrived at the date and hour appointed; they were received at the Astana of the Sultan of Perak by the Sultan, the Resident-General and all the Perak Chiefs. There was a Guard of Honour in attendance and a salute of seventeen guns was fired in the Sultan's honour....

Owing to a very heavy storm of rain, it was impossible to hold the opening meeting of the Council on the 13th, as had been intended, but at 10.30 am on the 14th, the four Sultans and their Chiefs, the four Residents and the Resident-General, met at the Astana of the Sultan of Perak, and there received the High Commissioner, who was in uniform and attended by his personal staff....

A great Fish Drive had been organised for the afternoon of the 14th, and everyone was there to enjoy it, but unfortunately this event was a failure, for there were very few fish, owing to the fact that the fenced enclosure into which they were to be driven was ill-chosen, being below a very deep pool wherein they naturally all took refuge....

Land sports were to have been held on the afternoon of the [next] day, but they were abandoned, because the assembled crowds of people seemed more inclined to amuse than to exert themselves. From 5 pm the Malay visitors called on the High Commissioner and joined a Garden Party at the

Residency where picturesque groups on the terraces were duly photographed. In the evening, the High Commissioner and his staff, the Resident-General, the Residents and all the Sultans were entertained at dinner at the Astana by the Sultan of Perak; and later the whole party visited a Malay theatre.

On the morning of the 16th, there was an extremely successful picnic at a waterfall about six miles from Kuala Kangsar, where most of the party participated in the amusement called *Menggelunchur*, which has been described elsewhere. Nearly all the Chiefs present joined in sliding down the rock and seemed to immensely enjoy an entertainment that was quite novel to visitors....

It had been arranged that Boat Races should be held on that afternoon but it was not possible to carry out this item of the programme as all the boats were occupied by people who were using them as dwellings.... In the evening there was a display of fireworks. As the last of these was fired before the time fixed on the programme for the commencement of the display, none of the guests were able to witness it....

It is an extraordinary thing in my experience for a Malay Sultan to have done what was accomplished by the Sultan of Perak on this occasion. For a man of his position, and with his traditions, to go through a week of ceaseless entertainment—his house filled by guests, many of whom required special consideration—to be kept up till all hours of the morning, and finally, with only three hours rest, to be up at 6.30 am to see the last of the High Commissioner, and then himself to accompany for several miles along their journey in opposite directions, first one and then the other of his two most distinguished guests, is a record which has never before been attempted, and probably will never be surpassed.... On each night of the week he gave a dinner party to between twenty and twenty-four guests; he used his carriages and horses perpetually, and he did everything in his power to make the visit of his distinguished guests as pleasant as possible for them.

More than this he gave them all souvenirs of their visit, and when they went to see the various entertainments provided for their amusement he made presents to the players, in accordance with Eastern custom, but in the names of his guests and not himself. It is a great satisfaction to know that the Sultan's efforts were thoroughly appreciated. Everyone has gone away delighted and loud in praise of the hospitality and thoughtful consideration of their host. The Sultan of Pahang made presents to the Sultan of Perak's sons and other young Rajas who helped their Ruler in the duties of entertainment....

The Sultan of Perak has told me, with pardonable satisfaction, that though for seven days many thousands of people had been collected in Kuala Kangsar from all parts of the country, such good feeling had prevailed that not even the children had quarrelled and there had been neither crime nor accident.

Frank Swettenham, unpublished Report on the Conference of FMS Malay Rulers at Kuala Kangsar in July 1897, enclosed with a despatch of 20 August 1897 (CO 273/229) from the High Commissioner to the Colonial Office.

72
Hari Raya at a Royal Capital

CHARLES TURNEY

Hari Raya *was the major public holiday of the Muslim year (see Passage 50 above) and seems to have been more elaborately celebrated than usual at Kuala Langat in 1894, perhaps because the ninetieth birthday of Sultan Abdul Samad fell in 1894 or thereabouts. He had reigned since 1857.*

C. H. A. Turney, the author of this passage and organizer of the celebrations to enliven 'the derary lives' of the local populace, was an 'old hand' who had come to Singapore in 1863 and to Selangor some twelve years later, where he held various posts. At this time he was the senior district officer in the Selangor civil service, stationed at the royal capital at Kuala Langat, and due to retire shortly after this festivity.

EARLY on the 18th all were astir, and the streets and Istana grounds were crowded with a cheerful and gaily bedecked crowd of Malays, intent on enjoyment. The Council chamber was specially prepared with carpets and a small dais with a cushion covered with yellow satin, for the reception of His Highness, and a piece of ground was laid out for sports, with a picturesque shed for the band, an octagaonal marquee for H.H. the Sultan and Rajas, etc., and at some distance, but commanding a view of the whole scene, a long shed for the ladies. The buildings were decorated with palms, ferns and flowers. The posts were entwined with yellow-and-white and yellow-and-red cloth, and pretty flags of various designs and sizes floated gaily over them. A large tastefully decorated frame, with a red-and-gold crown adorning it, and with the word 'Welcome' in large English capitals, in gold, on a blue ground, hung over the entrance, and was really a piece of work which did credit to the Raja Muda, who was the author of it. The school-boys were in various costumes; those from Telok were designated by a blue belt from shoulder to waist, those at Bandar in yellow (royal colours) and the Jugra contingent wore blue-and-red. The sports were really intended for the school-boys, and they knew it, judging from their eager happy faces. One lot of big boys from Permatang Pasir, who attended the Jugra School, were very picturesquely dressed with red Scotch caps, white jackets with blue-and-red sashes, white knickerbockers and scarlet gaiters. They had a band-master, dressed in fancy blue-and-yellow uniform, who put the boys through a drill. A violin accompanied them, playing well-known Malayan tunes, and also, it is funny to say, an adaptation of a good many modern popular English tunes, like 'Ring the bell, Watchman', 'Wait till the clouds roll by' and snatches from 'Pinafore', with verses in Malay composed for the occasion. . . .

The programme, arranged by the Raja Muda, was the Levee at 8 a.m., and Sports at 9, and it was anything but a short one.

The band was ensconced on one side of the Council Chamber, to the right, and a guard of honour awaited the advent of the Sultan.

At 8 a.m., precisely, the Sultan arrived, the band played a bar of the Selangor March, and as he passed through the ranks the guard presented arms. His Highness seated himself in front of the small yellow satin covered dais, looking quite pleased and cheerful, and the Rajas and Chiefs trooped in and sat down in file inside and along the rails of the Council Chamber. The crowd around the building was immense, and all were anxious to witness a ceremonial not carried out in so formal a manner for long years....

After the reception a prayer was recited and a blessing pronounced by the aged and venerable Imam of Permatang, Kuala Selangor.

The guard retired and the band marched away to the scene of the Sports, playing gaily, and followed by a large and appreciative crowd.

His Highness followed immediately after, and seated in his marquee, surrounded by his children and grandchildren, and his Rajas and Chiefs, entered with zest into the spirit of all the fun, and was full of smiles and laughter.

There were 23 events on the programme, the prizes varying from $2 to $10, the details of which would be monotonous to describe; the result of one of which, however, gave the Sultan and Raja Muda great pleasure. This was the tug-of-war between the police and crews of ships combined, against local wood-cutters. It remains for me to say the latter carried the day, to my surprise and the great joy of the Langat Malays.

One of the events was a very novel one, and that was an arithmetic race. The boys were handicapped by age and size, a slate and pencil was given to each competitor, two black boards were placed on the course with a small easy sum on each, and as the boys came up they had to copy and make up the sum, and run in to the winning post. The boy who came in first with one of the two sums correctly worked obtained the prize.

The feast after the sports was a hearty and generous one to which all resorted, and the three buffaloes (gifts of the Sultan) and some $400 worth of other good things [given] by the Raja Muda went a long way to satisfy the cravings of the hungry and thirsty crowd.

The day was a royal one, with just a little rain in the morning to keep down the dust, and cloudy and cool the whole of the rest of it. The band was attractive and gave zest to the proceedings, and the large concourse of people numbering, I should say, over 2,000, went away pleased with the amusement afforded to them for once in their dreary lives.

C. H. A. Turney, District Officer Kuala Langat, Monthly Report for April 1894, published in *Selangor Government Gazette*, 1894, pp. 340–1.

73
GCVO Week at Kuala Kangsar

THOMAS FOX

It became normal practice for the British Crown to honour a Malay Ruler with the award of the KCMG (Knight Commander of the Order of St. Michael and St. George); a tolerably successful British Resident might expect to become a Companion (CMG) of the same order. Investitures were thus not infrequent events. But Sultan Idris of Perak, who reigned from 1887 to 1916, was recognized to be the most outstanding of the Malay Rulers of the colonial period. Towards the end of his life, in 1913, he was awarded the GCVO (Knight Grand Cross of the Victorian Order), the highest award ever given to a Malay Sultan.

The investiture, at Kuala Kangsar in September 1913, entailed ceremonies which lasted a week and had a touch of historical pageant, as the British officials who organized the celebrations launched into some rather doubtful efforts at recreating the past (in a situation which had no Malay precedent). The Sultan readily fell in with this—there was an element of flamboyance in his character—but also made it plain that as a devout Muslim he found the austere dedication of a new State mosque to be the most important event of the week. This passage is taken from a booklet, written to record the week's ceremonies, by the then editor of the Times of Malaya; *the modern reader will discern some journalistic purple patches.*

THE arrangement of the programme was a matter of extreme delicacy, requiring the exercise of profound commonsense, considerable experience and an infinite amount of tact, to avoid the introduction of any element that could be interpreted as likely to clash with Malayan customs and traditions. The idea broadly was to convey the impression that the insignia had come direct from the King of England to the Sultan in his Palace at Kuala Kangsar. To create this impression it was essential that the insignia should be conveyed from some part of the river to a spot of Kuala Kangsar, and from there to the Astana Nagara. Adhering to a strictly correct line it would have been necessary to sail up the river, but the exigencies of tide and current could only be circumvented by sailing down.

It was at first intended to make the celebrations even a little more elaborate than they actually were. The original suggestions allowed for an attack by rebel spearmen on the party bringing the insignia, this being an acknowledgement of past fighting days, before the peaceful settlement of the country when life was held cheap and death faced every hour of the day, when rapine and murder were rife and progress stifled—days of semi-barbarism. This idea was eventually dropped.... Details innumerable had to be carfully considered and settled.... how many Malay School children should be invited? How many elephants would be required? How many lamps would have to be utilised for the illuminations?.... Mr H. Berkeley, the District Officer, Upper Perak, stated that he would be coming from his native fastness with a following.... 47 elephants, 25 buffaloes, 87 school boys, 65 actors, 24 cooks and over 100 men to help....

The spot selected for the start of the river procession was Enggor Pontoon Bridge, as being the most convenient. The barge was constructed of ordinary planking, supported by crossed bamboo poles, laid about six inches apart. There was a large canopied saloon with smaller saloons in front and behind.... The responsibility for conveying the insignia from the bridge to the High Commissioner's Lodge lay with Captain Oliver, who shortly before two o'clock arrived at the bridge, and accompanied by Raja Chulan and Raja Haji Abubakar stepped on board. Then the signal was given; the barge was loosened from its moorings and silently, but with a due sense of their responsibilities, the bargemen plied their poles and the State craft moved gracefully down the river. Around were scattered numerous small boats, gaily bedecked in colour schemes that left nothing to the imagination; they were formed in procession to accompany the main party.

Meanwhile large crowds had gathered on the right bank of the river at a spot opposite Saga. Elephants were already assembled at the other side. A great number of elephants from Upper Perak had never known a metalled road, seen a horse or heard a motor car, and on this account there was some uneasiness lest they should cause trouble. Thanks to the efforts of Mr Berkeley, who was singularly untiring, the elephants behaved themselves with the utmost propriety.

The ceremonies took place in the Throne Room of the Astana Nagara, and were witnessed by a number of Europeans and the Members of the State Council. The Throne Room is a little over a hundred feet in length and less than half that in breadth, magnificently upholstered in the Royal gold of the State, and richly laid with carpets that had obviously at one time reposed in some London warehouse. The Throne itself faces east and the light is thrown upon it through five open windows on each side. There is a modern touch about the interior, altogether out of keeping with the Oriental life and the Oriental dress.

Those presenting the addresses had to enter at the far end of the room from the balcony and in compliance with the force of Royal power and dignity, obsequiously proceed to within a few feet of the Throne; the occasion linked up the hands of the beautiful and the bizarre, the grave and the naive, a spectacle thoroughly Reinhardtic in character....

The white dusty road to the Astana Nagara after one o'clock was devoid of vehicles and given entirely over to pedestrian traffic. Europeans, Malays, Chinese and Indians alike trudged over the two miles in the midday heat of a garish Malayan day.

Pretty ladies, attired in cool white, and accompanied by white trousered black coated men, wearily tramped along the sunlit road; the Malay, like the Chief of Scottish historical story, in silence strode before the patient, happy, dusky better half, by whose side cheerily trotted little Malay toddlers; the Chinamen plodded steadily on, following each other at a distance of about a couple of yards, and the Indians came in close knit groups of purples, mauves and orange.... During the hours of waiting people disported themselves in various ways—wandering through the camp at the foot of Bukit Nagara, picnicing in *al fresco* fashion, and generally engaging in social intercourse.

At 3 p.m., the A.D.C. to the High Commissioner was conducted by the Raja Muda and Chiefs and Penghulus, on elephants, from the High Commissioner's Lodge to the Astana by the long route. The procession consisted of a Malay band, followed by Malay school children, accompanied by their teachers and carrying banners aloft. Then followed 200 Malays, armed with krisis, and 200 armed with spears, warriors all, by their presence symbolising the physical force that surrounds the Throne. The Kathis and Hajis followed, attired in full orders, and touching the whole with a note of solemnity and quiet dignity. Then lumbering slowly came some seventy huge elephants, under the superintendence of the D.O. Upper Perak, with spearmen riding by their side as much for effect as from any apprehension of trouble, although Mr Berkeley had to appease the leader by placing his hand on one of its tusks while it covered the greater part of the road. Bringing up the rear marched four jogan [standard bearers] and 200 Malays.

T. Fox, *The G.C.V.O. Week: An Account of the Celebrations at Kuala Kangsar from September 21st to September 28th, 1913, to mark the presentation to His Highness the Sultan of Perak of the Insignia of the G.C.V.O.*, The Times of Malaya Press, Ipoh, 1914, pp. 5–41.

74
A Malay Marriage

RICHARD SIDNEY

The elaborate traditional ceremonies of a Malay wedding, many of them of pre-Islamic origin, were suited only to the stateliness of royal courts and the gentle tempo of nineteenth-century Malay village life. For various reasons the trend in this century has been towards simpler observances. But the Malay wedding may still include the bersanding, *which serves much the same social purpose as the European wedding reception. The young couple sit silent and impassive while a host of wedding guests are gathered to pay their respects and enjoy the hospitality of the bride's family.*

Sidney here catches the contrasts in lifestyle of the growing Malay bourgeoisie of the towns in the inter-war period. From his description it appears that the bride's family were—perhaps to offset the 'incredible expenses'—combining the wedding with a coming-of-age ceremony of a young son, marked by public reading of the Koran and formal respects to his teacher.

MY car is waiting and all is ready to take us from this town to Port Swettenham, where the marriage is to be performed. We are waiting for Hamid. Soon I espy, coming down my drive, a gorgeous figure, who turns out to be Hamid dressed as a Malay. It is necessary to emphasise this, because he is one of the rising generation who much prefers the comfort of European clothes and their convenience to the more picturesque garments of his own countrymen. (I have seen Hamid struggling with a recalcitrant and heavy silk *sarong* and cursing the while that his countrymen demanded this curious dress of him. On that occasion he was going out to visit some relations of his, the chief of whom was a highly respected police officer. When Hamid had succeeded in doing himself up to what should be the satisfaction of all Malays, imagine his disgruntlement when his uncle turned up in an American motor-car and in the very latest European clothes!) ...

We found ourselves approaching a gaily decorated series of houses, and this was where the marriage was to take place. It was early in the afternoon and the bright sunlight danced on the flags of two *attap* huts, which had been specially built for the occasion, and on the bright *sarongs* and bajus of those Malays who were getting ready to welcome us. It is quite useless ... to attempt to describe the variegated costumes of the Malays. I saw every combination possible without there being any semblance of clash in the *bajus* and *sarongs* worn by the various members of the wedding party. One boy was clad entirely in black and had a vivid red *sarong* to throw it off. Another had a *baju* of red and a *sarong* of the brightest green, and trousers of a different shade again; and so it went on, and one longed for the time when colour-photography was possible.

The reader must imagine a small side road, on one side of which are fields, and on the other four small Malay houses, not of the good bungalow type in which Europeans live, nor of the usual Malay hut type, but a cross between the two. In front of these—for they are some way back from the road—is an open space on which have been specially built two huts so as to house the guests for the ceremony. One hut is large, and will take perhaps a hundred people sitting down on the floor. It is raised six feet from the ground and is approached by three planks, which make very awkward steps. It is thatched with the usual *attap* thatching, and for windows there are coloured flags. All around on the floor squat Malays, who are listening, perhaps, to the young Malay as he chants the Koran, or are talking among themselves and wondering whether the marriage will be successful, or what will be the price of rubber, etc. I was allowed to invade this hut and to take photographs therein, and it seemed that more interest was caused by my operations than by the legitimate droning which was proceeding at one end of the hut. I could only glimpse a picture of the young boy with his back to me, but could see that he was clad in gorgeous garments, and could hear him monotonously reciting the passages from the holy book of Islam. Outside this hut all was bustle and gaiety, and the preparation of food and the washing up of plates was perhaps one of the most noticeable features. The other place was more like a tent, and was not raised from the ground. It had tables and chairs in it, and we were invited to sit down and take tea and eat cakes, such as

A wedding gift on the hoof in the mid-1870s. The wedding guests look hungry and the gift apprehensive—both with good reason, from Isabella L. Bird, *The Golden Chersonese and the Way Thither*, John Murray, London, 1883.

are prepeared for *Hari Raya*. . . . There was no objection to my taking photographs, though I found it impossible to get a good one of the bride. She, with all the other womenfolk, was occupying a much shut-in house, and only occasionally did we get glimpses into this. Again there was the brightness of the *sarongs*, and the charm of the women's features, with their heavy gold ornaments, worn in honour of the marriage feast.

The ceremony took a long time, and in fact we saw little of it though we were there for two hours. Apparently the bridegroom himself has to make a pilgrimage through the town and then to come and claim his bride. For the first time they meet each other. Later they will sit side by side on a specially decorated bed and there receive the homage from relatives and friends. When this is over the bridegroom, curiously attired, must feast with his friends at a dinner which will consist, perhaps, of twenty or more courses! (And all this annoyed Hamid very much. The expenses, he told me, were almost incredible, especially as the bride's father had to provide bedding and cushions and bedsteads, etc., for the wedded pair.)

The chanting of the Koran being finished, we saw the young Malay boy taken out, and with a procession of elders make his way on the shoulders of another boy to the house of the priest, there to do obeisance. With him were ceremonial and very decorative boxes, which looked at first like large wedding cakes but which were actually models of a mosque. The last that we saw of the marriage was in the town itself, when a large crowd followed the procession of the bridegroom before he could come to claim his bride.

R. J. H. Sidney, *Malay Land*, Cecil Palmer, London, 1926, pp. 47–55.

75
Malay Music-making

DANIEL VAUGHAN

We have seen (Passage 49) that Vaughan's many accomplishments included proficiency as a musician. Malay and Western music are closer in their basic harmonies than say Chinese music. Europeans observed (for example, James Low in Perak—see Passage 35 above) that there was a mutual enjoyment in each other's music, and on occasion (see Passages 72 and 76 below) Malay musicians used instruments and played popular tunes of the West.

THE Malays are exceedingly fond of music and many have acquired a tolerable knowledge of the violin and play their national tunes and many European airs correctly; the drum appears to be the only native instrument, and for hours will a party of Malays amuse themselves by reciting verses accompanied by its monotonous tones.

On the violin they will execute by ear all their own tunes and English jigs and Portuguese fandangoes and will dance to the tunes with as much spirit as an Englishman at a fair; reels and jigs they manage well and will go through the figures of a quadrille tolerably; it must be stated that the stolid country Malay seldom indulges in such amusements, those that reside near the Town and the Jawibukans* are fond of imitating their European neighbours....

The airs are plaintive and consist of numerous semitones; they are exceedingly pleasing and excel in sweetness the Chinese and Indian melodies. At festivals the drums and cymbals are the usual instruments employed, shrill pipes resembling the clarionet are introduced but they are always played on by natives of India. Although Malayan music abounds in semitones they are not numerous enough to destroy all traces of a regular melody as is so remarkable in Chinese music. Malayan tunes are capable of being easily adapted to European music, and a musician of Singapore has set several for

*Abbreviated form of Jawi-peranakan, persons of mixed Malay and Indian Muslim parentage, in Georgetown, Penang.

the piano and other instruments, and they prove as popular as Jullien's productions.

Malays are so fond of musical sounds that they very ingeniously provide themselves with an instrument or rather instruments which furnish them with unceasing strains of harmony. A branch or sprig of bamboo is fixed on a pivot at the end of a long pole, which is stuck in the ground with the sprig in the air, the latter is set in motion by the wind and produces a clear full toned note, the size of the branch alters the tone, and when a number of such contrivances are erected of different sizes they produce a chord of musical notes nearly perfect; their taste and ear is so correct that discordant notes are seldom heard; the sounds of these whirligigs are heard fully half a mile off, and as the breeze lessens or increases so the notes swell or subside, pleasing the ear; each village or campong has its Aeolian band, so that the traveller catches the notes of the one he approaches as he loses the strains of the one left behind.

Their love of music is so deep set that they may be induced to work hard for several hours without flagging or uttering a word of dissatisfaction, if allowed to gratify their passion for music by singing while at work or having a musical instrument played to them. A boat's crew will pull seventy miles, with scarcely any cessation, if allowed to sing, and the notes of the fiddle will keep Malay lascars [seamen] at work for several days and nights without a murmur, and during that period they will rest merely to get their meals and sleep only for an hour or two at night; if it is not usual to have fiddles on board country ships manned with Malays the experiment ought to be tried, and the writer feels sure the plan would be profitable. While paddling in their native boats they sing and beat time to the measure by striking their paddles against the gunwale of the boat, occasionally the melody ceases whilst the regular beat of the paddles on the boat is prolonged to assist in keeping the paddlers together; they also vary the time by altering the method of beating—thus, they strike once, rest awhile and then strike twice rapidly before each stroke, or they beat three times at regular intervals, rest for a few seconds and then pull, or they paddle very fast striking the gunwale after each stroke of the paddle, or they paddle three times and then stop for a few seconds; one method is exceedingly pretty, it is called burong turbang or birds flying, it is done by paddling very rapidly with or without striking the boat after the stroke, and by each puller throwing the water he brings up on his paddle into the air away from the boat so as not to sprinkle those in the boat. To their musical taste may be attributed their methodical way of paddling and keeping time; unmusical people like the Chinese and some of the tribes of Hindustan care little about pulling together in a boat; by doing so they not only make it inconvenient for themselves but also retard considerably the progress of the boat.

J. D. Vaughan, 'Notes on the Malays of Pinang and Province Wellesley', *Journal of the Indian Archipelago*, New Series, Vol. 2, 1857, pp. 135–7.

76
A Malay Production of 'Hamlet'

RICHARD SIDNEY

In his work as a schoolmaster Richard Sidney (Passages 16, 27, and 74 above) produced, and played the leading roles in, school productions of Shakespeare. Hence he took a professional interest in contemporary Malay theatre ('opera' is rather a misnomer). The leading Malay touring company at this time was the 'Wayang Kassim' (Kassim's theatre company) and Sidney is probably here describing one of its productions.

MALAY opera is very different from a *ronggeng*,* and worth describing in some detail, because it is a form of amusement which appeals so intensely to many Eurasians and Chinese, as well as Malays, throughout the Peninsula. The reader must imagine either a long wooden hut, a barn-like structure with an *attap* roof, or a very long tent into which is crowded the audience, the stage and everything else. The stage itself, and all its apparatus, does not take up an enormous amount of room, and the audience will sit

*Malay dance; see Passage 77.

on chairs; the chairs for the stalls being of a better quality than those of the 'pit', which is just behind. Occasionally there are raised stands at the back which serve as a 'gallery'. One will make one's way through an aisle down the middle of the theatre and take one's seat, preferably, in the third or fourth row. From here one will be able to note the various details not only of the audience but also of the actors and orchestra, etc. We will suppose that the play has not yet begun, but that there has been handed to us a programme giving information about what is coming. The programme is a long, thin, pink paper, partly in English. After a preliminary flourish it begins:

THE ROYAL FAMOUS BANGSAWAN IN S.S. & F.M.S.
FOR MORE FEW NIGHTS ONLY!

Then it will warn you in heavy type to

LOOK OUT!!!

After this it continues:

> The actress will act in the leading parts of the play and the clown will keep you roaring with laughter.

(This was no understatement, for there were two clowns and they certainly did keep us very much amused, or at any rate that portion of the audience which could properly appreciate their jokes.)

Glancing again at the programme one reads that the company will stage a play entitled *Prince Hamlet*, and thereafter follows a synopsis of the play in Malay. Before continuing with the play itself let us look round and get our bearings, for it seems unlikely that the curtain will go up for a few minutes.

The audience is very talkative, smoking goes on everywhere and gradually the place fills up, so that the stalls, at any rate, are full. Just in front of us is an orchestra which has much music before it and promises to be very amusing. I can catch sight of a violincello, a piano, two violins, a flute, a clarionet, and a jazz combination consisting of cymbals, drums, etc. They are already tuning up, and very soon the overture breaks out. Meanwhile quite near me on the left is a large refreshment-stall, with ice and all the various appurtenances for looking after the appetites of those who will spend the next four hours in this tent or barn.

The Malay opera seems to have existed for some thirty years, and used to be known as a *wayang*. The word 'opera', however, was adopted about ten years ago, and is probably due to many of the modern tendencies which one can see throughout Malaya. It will be noticed throughout that there is an attempt to imitate the European method of doing things.... And now the curtain rolls up and we are confronted with a very passable imitation of English scenery, though all is silence. One might have expected the ordinary first scene which Shakespeare wrote for *Hamlet*; but this would not do for the Malays. Sleeping on a couch at the back of the stage lies an old gentleman. Everything is in dumb show and no word is spoken, and soon we realise that we are being shown the death of the old King. As soon as the murder has taken place—by a mouthing individual who turns out to be the future King—the whole company troops on to the stage and wails! The Queen had been watching her husband's murder from the side of the stage and it was her shriek that brought on the company. No longer is it dumb show—in fact, there is almost too much talking, though we have not yet got near the first actual scene of *Hamlet*. It must be understood that in a Malay opera no one carries on with the ordinary tale until he or she has announced to the audience the part which he or she is playing and what is to be done. There is thus a constant interruption of the action. Hamlet, however, is easily distinguishable, and he has evidently been chosen for the quality of his speaking voice.

The play now begins to follow Shakespeare's version in its main outline, but we do not have to wait for humour until the gravedigger's scene, for it begins as soon as the first clown (the sentry) meets the ghost! This is really funny! A telephone has been introduced into the castle grounds at Elsinore (though I saw this play actually in 1923 the producer must have had a foreknowledge of the work which Sir Barry Jackson was going to do in 1925), by means of which the comic sentry tries his best to call up the rest of the guard. The ghost, however, will stand none of this nonsense, and continually frightens the sentry, so much that he falls down on the stage, kicking up his legs in the air. (Loud laughter and cheers from the audience.)

The play continues to follow Shakespeare in its main outlines, though interpolating here and there scenes and explanations which make it all very long, so that by midnight Polonius has only just been killed and there is no sign yet of Ophelia's madness.

Everything, as I have said before, is a close imitation of the methods used in the West, though the imitation, unfortunately, is very bad. The scenery consists of flats, and of backcloths which roll up and down—they cannot be pulled up because there is no room above the stage. The scenes are badly painted, faulty in perspective, and bear very little relation to what is going on in front. A street scene of a town which we all know very well will be let down in the middle of a play whose authors had never heard of Malaya! Occasionally, however, there is something almost beautiful, and the illusion, for a short time, is complete. The dresses must cost a great deal of money—many being of gorgeous velvet, and almost too hot to think about so far as this climate is concerned. The make-up is patchy, and little chance for doing this well is given to the actors, because not only is the stage lighted quite unevenly but there are large lights in the audience throughout the performance. I once noticed a girl who was made up with a beautiful pink and white face but who still had black legs!

I have said that characters announce what they are going to do and who they are; and very often they will sing a song in Malay but with the words set to old English or American popular songs which we heard perhaps in the early days of this century. *The Belle of New York* is a very favourite opera from which tunes have been taken. It does not matter either whether all the parts of the orchestra are working in harmony, nor particularly whether the person on the stage is singing the same tune as the various parts of the orchestra are trying to play! Sometimes the whole song is a series of discords, and nothing can contrast more with the Chinese theatre than this particular side of the music.

R. J. H. Sidney, *Malay Land*, Cecil Palmer, London, 1926, pp. 141–5.

77
Malay Dancing

FREDERICK WELD AND OTHERS

Malay dancing nowadays suggests the ronggeng, *and indeed the visitor is more likely to see a performance of this traditional dance, transmuted into a rather commercial modern style, than any other. It came originally from Java where it derived from dancing in Hindu temples in pre-Islamic times. But there were other formal dances performed at Malay courts, and country dances in the villages, though both have almost disappeared now, owing to changing fashion and partly because modern culture, influenced by Islam, disapproves of it and accords low status to the professional female dancer. 'The* ronggeng *is not trained for the purpose; she only takes to dancing when her reputation is beyond the risk of any further injury' is a twentieth-century verdict, though perhaps not entirely fair. Even without long years of training the* ronggeng *dancer is a very skilled performer who may sing and act as well as dance. Islam, although not sympathetic to public dancing of the two sexes, introduced its own ceremonial dances of Arabian origin as a religious performance.*

The characteristics of Malay dancing of all kinds are the importance of 'the undulations of the fingers and arms, and the swaying of the body' in addition to the steps of the feet (there are different Malay terms for each of the three movements); the entertainment of the audience rather than the enjoyment of the performers is the main consideration; and there is often an element of competition or repartee—it is a contest as much as an art form.

Nineteenth-century Country Dances

AT night sitting amongst palm and banana and other rich tropical foliage intermingled with scarves and flags and quaint devices on the banks of a mountain river lit by a full moon in a cloudless sky we witnessed a series of dances and performances and listened to the chants of the wild Sakeis, the Malays, the Menangkabau men and the Chinese—we had scarf dances and shawl dances and saucer dances and a monkey dance and besides singing,

Malay and Chinese instrumental accompaniments, all the performers being men....

* * *

Sergeant Syed said the Police wanted to come to my house in the evening to 'Main Boriah'. At 8 p.m. a procession with torches came to the house, the men were dressed in white with scarves and caps ornamented with tinsel, and one of them sang a long song, written out, about the affairs and people of Kajang, the song was sung to a very pretty tune and the chorus was a good swinging one, and all hands joined in heartily. After this several dances were performed, and everyone seemed to enjoy himself thoroughly.

At a Royal Court in 1875

When we entered, we saw seated on a large carpet in the middle of the Hall, four girls, two of them about 18 and two about 11 years old, all beautifully dressed in silk and cloth of gold.

On their heads they each wore a large and curious but very pretty ornament, made principally of gold—a sort of square flower garden where all the flowers were gold, but of delicate workmanship, trembling and glittering with every movement of the wearer....

All four dancers were dressed alike, except that in the elder girls, the body of the dress, tight fitting and shewing the figure to the greatest advantage, was white, with a cloth of gold handkerchief tied round it under the arms and fastened in front, whilst in the case of the two younger, the body was of the same stuff as the rest of the dress. Their feet of course were bare....

From the elaborate and vehement execution of the players [of the orchestra], and the want of regular time in the music, I judged, and rightly, that we had entered as the ouverture began. During its performance, the dancers sat leaning forward and hiding their faces as I have described, but when it concluded, and without any break, the music changed into the regular time for dancing, the four girls dropped their fans, raised their hands in the act of 'Sambah' or homage, and then began the nautch by swaying their bodies and slowly waving their arms and hands in the most graceful movements, making much and effective use all the while of the scarf hanging from their belts....

They danced 5 or 6 dances, each lasting quite half an hour, with materially different figures and time in the music....

The last dance, symbolical of war, was perhaps the best, the music being much faster almost inspiriting and the movement of the dancers more free and even abandoned. For the latter half of the dance they each had a wand, to represent a sword, bound with three rings of burnished gold which glittered in the light like precious stones.

This nautch, which began soberly, like the others, grew to a Bacchante revel until the dancers were, or pretended to be, possessed by the Spirit of Dancing 'hantu menari' as they called it, and leaving the Hall for a moment to smear their fingers with a fragrant oil, they returned, and the two eldest, striking at each other with their wands seemed inclined to turn the symbolical into a real battle. They were however, after some trouble, caught by four or five women, who felt what magic wands could be made to do, and carried forcibly out of the Hall. The two younger girls, who looked as if they too would like to be possessed but did not know how to do it, were easily caught and removed.

Muslim Dancing

Moslem influence has introduced the *hathrah*, or catechismal dance, a form of dissipation that any pious *haji* can safely patronise....

Usually the *hathrah* is danced by a long line of boys who sing, sway, and prostrate themselves before the venerable pundit who instructs them in religious chants. The words are largely Arabic; the sentiment is religious; the professed object is to glorify God; the cost of the entertainment is met by a public subscription or by the generosity of the patron who gives it. When it opens, the boys are seen seated on a mat in front of their catechist who burns incense and exhorts them to devotion. They then rise and repeat a long chant, accompanying the words with certain slow, graceful and rhythmical movements, and ending the performance by falling prostrate before their teacher in the humble attitude of prayer. The general effect is pleasing. The uniformity of the costumes, the rhythmical unity of the dancing, the sweet boyish voices intoning the solemn Arabic words in the still night air, the softness of the light, the reverential gravity of everyone; these things combine to make the European spectator realise the possibilities of the religious dance.

Ronggeng

This dance had several novel features: the youths and girls continued to dance in two lines facing but not touching each other, but emphasis was laid on step dancing, and the movement of hands and arms were restricted and of secondary importance. The music followed western form, with western scales and a new stringed instrument, a violin was introduced from Europe, though the barrel-drum and the hanging gong were retained. If a violin was not available, a bamboo flute called *seruling*, which could play a western scale, was used instead.

At some period, following the Portuguese occupation of Malacca, unmarried Malay girls were banished by public opinion from the dance floor of at least the west coast states, and professional dancing girls, who had long been an elegant feature of Javanese cultural life, took their place....

* * *

Some good performers are paid not less than five hundred dollars for a night's show; while others can command only fifty dollars, or even less.... The difference is mainly due to the type of women who take part. If they happen to be amiable girls between the ages of fourteen and twenty the demand for their services will be greater, and the payment considerably more. When they get older the demand decreases, and so does the payment. You will find the younger ladies in the towns and those whose charms have faded in the smaller *kampongs*....

Three young and stylishly dressed Malays were dancing opposite to their female partners, and one watched them for perhaps half-an-hour and the young men got tired considerably before the ladies, who could not be outwitted in any step suggested by the male.

It is no uncommon thing for Europeans themselves to take part in these dances, and, in fact, there is nothing very difficult about the steps, though he would be a bold man who tried to tire out a professional!

The crescendo of the *ronggeng*, from Donald Moore, *Far Eastern Agent or The Diary of an Eastern Nobody*, Hodder and Stoughton, London, 1953.

F. A. Weld (reporting on a visit to Ulu Langat) in a despatch dated 19 July 1883; and monthly report of the District Officer, Ulu Langat, in *Selangor Government Gazette,* 1891, p. 834, on country dances; F. A. Swettenham, 'A Malay Nautch', *Journal of the Straits Branch of the Royal Asiatic Society*, No. 2, pp. 9–12, on a visit to Pekan, royal capital of Pahang, in 1875; R. J. Wilkinson, 'Malay Amusements', *Papers on Malay Subjects: Life and Customs*, Part III, 1910, pp. 28–9, on the *hathrah*; Mubin Sheppard, *Taman Indera: Malay Decorative Arts and Pastimes*, Oxford University Press, Kuala Lumpur, 1972, p. 89; and R. J. H. Sidney, *Malay Land*, Cecil Palmer, London, 1926, p. 139–41, on the *ronggeng*.

78
Kites and Tops in Trengganu

NOEL REES

Owing to its comparative inaccessibility until very recent years, Trengganu retained its overwhelmingly Malay character and way of life longer than most other parts of the Peninsula.

Noel Rees was a member of the Education Service, who served in Trengganu in the 1920s and had the opportunity of observing some of its traditional Malay amusements, in this case kite-flying and top-spinning. As with so many of these competitive sports (see Passage 85 below), there was much betting on the outcome. Rees was a celebrated raconteur.

WHEN I first went to Terengganu I was told that it was an earthly paradise, and it was. It was a lovely, quiet country, full of quiet, friendly people.

I shall start by reading part of one of the Terengganu Government Proclamations. It was written in Malay, in the Arabic script, and had been translated into English by the then British Adviser....

Proclamation No. 17 read as follows: 'Be it known unto all men that owing to the danger of injury to persons using the public road, on or after the date of this proclamation, the practice of flying kites, large or small, with tails or without, is strictly forbidden in this town of Kuala Terengganu. Any person found disobeying this proclamation may have his kite confiscated by the police and taken to the police station.'

There was only one road with hardly any traffic on it except the occasional bullock cart. Kite flying is hard to describe but they are enormous and it needs two persons to get them off the ground. They have village competitions to see whose kite can destroy whose! They are beautifully coloured and in all kinds of shapes and sizes. They put cut glass on the end of the kite to try and cut the string of the other kite. Sometimes the string may be more than half a mile long. This takes place in the monsoon season when the padi planting is over and there is nothing else to do. It is a game for strong old men, not for kids and crowds of people come out to witness the competitions.

Another favourite pastime was *top spinning*. This is also a game played by strong middle-aged men. The tops used are very heavy, weighing almost 2 pounds in weight. They are hurled on to a hard patch of ground and then lifted with a scoop and put on a kind of pewter plate and they go on and on spinning. Some would spin for getting on for an hour without wobbling. Now about three houses down from mine there was a wiry old Malay called Wan Hawan. I knew him to be the most famous top-spinner for miles around. I went to see him to ask him if it was true that these tops could spin for an hour. He laughed and told me that was nonsense. He had a top that would only start to wobble after an hour and a half! Then he asked if I would like to see it. He went into his house and brought out a bit of cloth and inside was this enormous top, flat and made of wood with a steel band around it and a very beautiful polished bit of brass at the top. It was a beautiful thing to look at. He asked if I wanted a demonstration which of course I did! He put the string around it, then lifting it a full arm length above his shoulder, he hurled the top and it fell on the hard ground. He took his scoop, scooped the top up and put it on the prepared plate on top

A spinning top launched into flight at the start of its gyrations, from *Pesta Official Souvenir Programme*, Information Services, Kuala Lumpur, 1956.

of the bamboo and there it was spinning. He suggested we had a cup of coffee and after an hour the top was still motionless on the plate and I remember seeing the sun glinting on its steel surface. After an hour and forty minutes, it was still going strong and I asked him what would happen if I touched it. He said it would take the skin off my fingers and break my wrist as if it were a dry twig. He was much feared because if he was participating in the village competition all the money would be on him. Betting was very common for top spinning and if it was known that Wan Hawan had a top, all the money would be on him. Now I had heard that he was a 'top doctor' and that he used wizardry and could influence a top whilst it was spinning. I had also heard that he could make a top wobble without touching it. 'Of course I can,' he said when I asked him. 'All I need to know is where the competition is taking place, at what time and on what day of the week. Provided I know where it is, say 56 miles to the North East, on a Friday afternoon at 3 o'clock.' I asked him what his extreme range was and he replied '80 miles!!' He was a very cunning fellow. He would attend all the contests. Once there was one in Kelantan. He was not directly engaged in it but he went there and sat under the trees chatting away. They were watching him to see on which top he would put his money, and then they would follow him! He sat back in a disinterested way but once the contest was under way, he took a 10 dollar note and slipped it into the hand of one of his accomplices, in such a way that people would see but at the same time trying to kid them that it wasn't. The accomplice walked off and put the 10 dollars on Top B. The news spread. Women would pawn their necklaces to get money for the top spinning. So the money poured into Top B. Now Wan Hawan had his money on Top A. The first top to wobble was the loser. So they sat there for an hour. On that occasion Top B started to wobble. He was cursing away but in actual fact he had money on Top A. He would wobble the favourite and the net profit would be 590 dollars to him. He was a great friend of mine and once I was drinking coffee with him in a coffee shop on the Kelantan border. He was on his way back from a contest. As we were talking a bus came round the corner, going past at tremendous speed, full of young men shouting and singing. The proprietor of the shop told us it was the Kelantan football team going to play Terengganu that afternoon. 'Call them back the idiots! The fools are blind, coming out from the North East on a Friday afternoon. If they are not beaten 3–0, then when I am gone, take the stone off my grave and hurl it into the sea from the rock of Cenering.' I went back that afternoon and there was great joy in the capital because Kelantan had been beaten 3–0!! That was absolutely true and when you live in the East you come to believe in things you would never believe in England!!

Noel Rees, 'Tall Tales from Terengganu', *Journal of the Malaysian Branch of the Royal Asiatic Society*, Vol. LXIII, Pt. 2, 1990, pp. 69–71.

79
Club Life in 1910 and 1972

HENRI FAUCONNIER AND ANON

In Kuala Lumpur the two leading clubs were the Selangor Club, founded in 1884, and the Lake Club, founded in 1890. The former had the nickname of 'the Spotted Dog' for reasons which Fauconnier explains.

In the first section of this passage Fauconnier, a French planter from an estate in the Kuala Selangor district, describes the monthly carousals of the planters who came in to town on the first Saturday of each month to draw cash from the bank for the payment of labourers' wages for the previous month. But before they returned to their estates, they met to have a drink together.

With the passing of the colonial regime the Lake Club, long a bastion of privilege with its membership confined to the more senior members of the European community, took on a new lease of life, expanding its activities and its membership. The tone of the 1972 notice indicates the organized vitality of its current existence.

1910

THE planters, who are fond of nicknames, had called this club 'the Spotted Dog'; for the reason that, unlike other clubs, which insist on an unimpeachable white skin, this one admitted members more or less tinged with a darker hue, Eurasians, and even pure Tamils. They seldom came, but they had the proud pleasure of paying their subscriptions.

The hair-dressing saloon was kept by a magnificent Hindu, assisted by several apprentices. On leaving the bank I often came in there to enjoy a little doze, while aerial scissors clicked hypnotically round my head, and deft fingers performed a massage of my skull, that looked like a conjuring trick. Under the gentle treatment I felt as if all my troubles were dispelled, my clarity of mind restored.

Soothed and perfumed from the hands of this magician, I made my way towards the bar. It was packed. The semicircular counter, edged by an unbroken line of lifted elbows, was inaccessible. But beyond it were a few little cane tables and chairs for those whose thirst was not so urgent. The company always arranged itself in geographical order, and I was familiar with the zone of occupation assigned to my district. But I had not taken three steps in that direction before I was observed, caught by many hands, forced into a chair with a large gin-sling in front of me that had been waiting for me since midday, and all glasses were lifted in my honour.... Flowerpet, unbuckling his leather belt, tried to make it into a crown for me, with great key rings dangling over my ears. Old Holmwood had hoisted himself on to a table, and there stood, with congested eyes and violet cheeks, trampling on broken glasses and brandishing a cane chair.

1972: Tennis by Night

A Barbecue at the Tennis Verandah has been arranged for SATURDAY 13th May to celebrate the official opening of the new lighting system.

It is hoped that players and other interested members will turn up and get to know each other at this informal get-together from 7.30 p.m. Should the weather be unfavourable, tables will be arranged in the Banquet Hall with table tennis as a consolation.

The lights will be turned on at 6.40 p.m. to deal with the dusk invasion by bugs, etc., and the first balls will be served at 7.30 p.m. by our two active tennis-playing members of the General Committee, Tan Sri Haji Nik Daud and Mr Sydney Woodhull. After the opening play, it is hoped to have some keen games for the rest of the evening.

A Word for the Ladies—Insect repellant might be handy to have around, in case of invasion during the first hour from 6.45 to 7.45 p.m.—after that, it's not so bad.

To assist with catering arrangements, can you please let us know if you're coming. Once here, we will have a table for you but you eat when you please, between 8.30 and 10 p.m.

1910: Henri Fauconnier, *The Soul of Malaya,* translated from the French by E. Sutton, Elkin Mathews and Marrot, London, 1931, pp. 189–91; *1972:* D. J. M. Tate, *The Lake Club 1890–1990: The Pursuit of Excellence,* Oxford University Press, Singapore, 1990, p. 125.

80
Sport in the 1890s—the Lighter Moments

JOHN RUSSELL AND OTHERS

Most of the following reports of sporting activities (and also Passages 80–83 below) come from the Selangor Journal, *which from 1892 to 1897 was the fortnightly carrier of news and local gossip to the European community of Selangor. Its editor was John Russell, the head of the government printing department whose presses printed the journal. But a number of other officials, including J. H. M. Robson, contributed news items, often anonymously. In 1896 Robson (see Passages 19, 69, and 81) left the government service to found Kuala Lumpur's first newspaper, the* Malay Mail*; and so the* Selangor Journal, *replaced by a more frequent purveyor of news, was wound up. It remains nonetheless, as the following passages show, a unique record of the outdoor and indoor amusements of the community for which it was written.*

Ladies Cricket

AN amusing cricket match was played on the ground near the Lake, on Wednesday afternoon, 20th September [1893], between a team of ladies, captained by Mrs Gordon, and a team of men under Mr Dougal. The men had to bowl and field with the left hand, and to bat left-handed (using both hands) with broomsticks. It was a pity that the ladies had had no previous practice, but they all did well.... The decisions of one of the umpires were received with much approval—especially from the ladies. It is sad to relate that, notwithstanding the heroic attempts of the ladies and their extraordinary exertions, they suffered defeat.... One concluding word of advice to the ladies: the grand rule of whist, silence, should also be observed while fielding.

Malay Football

Some three days a week the Bank end of the Parade Ground is used by the Malays as a practice ground for football.... At present the players are Malays only. An amusing half hour may be spent in watching them. The natural gravity and dignity of the Malay is easily noticeable. No preliminary horseplay or turning of somersaults as amongst English school-boys can be seen. After a lot of talking, which is shared in equally by all the players, the ball is started. The game reminds one of the descriptions one reads of ladies' football matches in England. The players make apologetic charges and stand around in picturesque attitudes. The full back may be seen stretched at full length smoking a cigarette whilst a mildly fierce battle is being waged near his opponents' goal. Should the ball happen to trickle his way he will come to life and spread himself around gracefully until the tide of conflict has once more rolled backwards. The costumes worn by the players are very 'chic'. Where Dolah got those nice boots and stockings from must remain a matter for conjecture. Perhaps his long-suffering tuan would a tale unfold? Joking apart, however, these fellows will soon get into a better style of play and help fill up the ranks in the regular football matches played by Europeans.

Walking

Last week [Mr Hancock] appeared among us again and announced by hand bill and otherwise that on Friday, the 8th [June 1894] he would walk four miles on the Parade Ground against eight competitors, taking up a fresh man at each half-mile. A wet afternoon, however, prevented this, so that the event was postponed till the Saturday, when Mr Hancock competed against seven stalwart Sikhs, 'picked men', all up to the standard of Bill Adams' famous heroes, and with the voluntary addition of our athletic Auditor [Mr Vane] and Mr Owen. The weather was all that would be desired and the ground fair, though the circuit was rather limited, the walkers having to do eight laps to a mile. Mr Hancock started with Sikh Wy, who managed to keep up a gentle trot all the time. Sikh Wogy did better, but 'broke' at any time Hancock pressed him. The same may be said of the others till the last, Oh Gourka, came. This man undoubtedly walked well and steadily in advance of Hancock, and only when pressed hard did he at all break. Hancock just captured him on the line.

At this stage Owen started, but after one lap that gentleman clearly showed that the pace was too much for him. Mr Vane, in his quarter, walked remarkably steadily and well, making the pace, but was also passed, Mr Hancock coming in swiftly and displaying a wonderful lot of energy at the end of his four miles (32 laps)....

It may interest some to know that Mr Hancock's record for an hour is 8 miles with 17 sec. to spare. In 1880 he won the Championship of the World, a challenge cup valued [at] £120, and having won the Championship of England three times in succession....

Turf Club Meetings

There were no professionals at the Turf Club in those days [1896]. The races were run by amateurs and it was like a big family party. Ladies did not do their own betting.

A steward, one of the British members, always came up to the Ladies' Gallery and asked if they would like to buy a ticket.

The Chartered Bank closed on Race Days and took all their clerks, etc. to the Turf Club and ran the betting part. There was no Totalisator.

At 5 p.m. when the last race was run people strolled about the Padang to get good views of all the dresses....

There was a native stand a short distance away from the club stand. Malay royalty and heads of the Chinese and Indian communities came to the European side, being members of the club and owners of racehorses.

Tea with ice cream was served. Ices were always made in those days in portable tubs specially made to contain freezing ingredients.

Ahtletic Sports

On Saturday, the 14th [October 1892], by the kindness of Mr Khoo Mah Lek, some athletic sports took place on the grounds adjoining the Selangor Chinese Club. The grounds are rather too small and only allowed a circular course of 100 yards. There was a long list of judges, handicappers, stewards, starters, timekeepers, etc., but most of them being competitors the duties pertaining to their posts were somewhat neglected. We give below a short account of the various events....

4. Breaking the pot blindfolded—12 competitors. This was a very amusing event, four earthen chatties being suspended from a horizontal bar and each competitor having to walk blindfolded 40 yards and break the chatty with a stick; very few accomplished the feat....

9. Eating biscuits—this was a very amusing competition and was won by Mat bin Brahim, who said afterwards that he will never enter into another competition of the same kind....

13. Climbing the greasy pole—this event fell through....

16. Duck-hunt—this event caused great excitement. The river being very swollen and running rapidly the ducks and their pursuers were quickly carried away to the opposite shore. Won by a smart Malay boy, who went up stream a little and allowed himself to be carried by the current to the other bank and there found one of the ducks.

17. Catching the greasy pig—This was rare fun for everybody except the pig, who was frequently under a heavy scrimmage. The winner of this obtained the price, not the pig; and after Mr Dennis had been consigned to his sack he

was heard giving vent to his disapproval of the proceedings by much grunting.

Golf

One of the major drawbacks to the Petaling Hills site was that it contained a disused Chinese graveyard, and although at a later date (it is said) many of the skeletons were disinterred and sent to China by descendants motivated by the traditional filial piety of their race, there was for many years afterwards (again, it is said) the diverting possibility that a relic in the shape of [a] human skull or bones might be dislodged by a player who wandered into remote corners and sought to extricate himself by the firm, rather than scientific, use of a niblick.

One of the early local rules [provided that] 'Graves, with their mounds and trenches, roads, paths, and any part of the course over which the turf has been removed must be treated as hazards [but] long grass is not a hazard'. This rule, the captain interpreted, 'means that you cannot ground your club in addressing the ball or move anything, however loose or dead it may be, when you find yourself in a grave....' The deep ravine ... was known as 'Hell'.

[Other rules were that] a ball lying in Yap Hon Chin's boundary ditch or in the buffalo compound near the second green may be lifted and dropped under a penalty of one stroke [and] a ball lying in temporary water may be lifted and dropped without penalty.

The caddies in those days were all Malays. This is said to have been due to the fact that the site of the Petaling Hill course had been at some time a Chinese graveyard, and Chinese boys were consequently very reluctant to venture there....

The caddies' competitions were something of an institution in the old days at Petaling Hill.... They would themselves, on the day of their competition, prepare a Malay curry tiffin for the members they served in conjunction with the clubhouse boys, after which they went out to do battle, each pair being accompanied by a member whose duty it was to mark the cards and see that the rules were properly observed.... There was often a good deal of betting on the results of these competitions in the old days and 'it has been known for a member's arithmetic to be suspect when one of his men won'.

John Russell, *Selangor Journal*, Vol. 2, pp. 8–9 (ladies cricket); Vol. 4, p. 110 (Malay footballers); Vol. 2, p. 324 (walking match); Vol. 2, p. 41 (athletic sports); Mrs E. Stratton Brown, 'Looking Back on Selangor in the Nineties', *Fifty Years of Progress 1904–1954, Malay Mail Supplement*, 1955 (race meetings); Selangor Golf Club, *Twelve under Fours—an Informal History of the Selangor Golf Club*, 1953, pp. 8–9, 14, 40, 48–9 (golf).

81
A Night Out

JOHN RUSSELL AND JOHN ROBSON

As with the preceding passage, some material here comes from the Selangor Journal; *the remainder is from the recollections of J. H. M. Robson (see Passage 69), who arrived in Selangor in 1889. The final sentence refers to the Sino-Japanese war of 1894. A sulky was a light two-wheeler carriage drawn by a single horse.*

After the Ball is Over

AN amusing story is told of two gentlemen travelling between Sungei Ujong and Selangor after the late race meeting [in June 1895]. It appears that these gentlemen left Seremban in a sulky during the small hours of the morning, after dancing vigorously at the Resident's Ball, and being fatigued with their exertions it is presumed they were dozing peacefully on the homeward journey. All went well until they arrived near the Beranang Police Station when, to their horror, they saw a strange beast approaching, looking as if it meant mischief, and having no weapon at hand, they sought safety in flight, followed by the beast, which stuck to them persistently for a couple of miles, though they are said to have done a record. During their run speculation was rife as to the nature of the animal in pursuit; one declared it must be an elephant, and the other thought it a tapir, but both were quite of one mind as to the advisability of getting away from it as quickly as possible. The party reached Semenyih Police Station greatly excited and sought the corporal's assistance to destroy or capture the daring creature who had chased them so persistently. The corporal marshalled his forces and was about to take the field when the pursuer turned up, and, instead of the dangerous animal represented, he proved to be a donkey belonging to one of Kajang's most prominent citizens....

Among other things [the owner] said 'One must have been to a race meeting in order to mistake the ears of a donkey for the horns of a tapir.' ... The only sufferer is the donkey, whose master declines to keep an animal that goes fooling around at night trying to frighten people; he is to be sent to Seremban, and placed under supervision.

Concerts

In the absence of a Town Hall, the Selangor Club was used for entertainments and travelling shows. The first film to be seen here was shown at the Selangor Club. It was a poor, flickering production. Smoking concerts were held at the Club to which ladies were not admitted. Some ladies resented this exclusion, so on one occasion two of them hid under the building in order to hear what was going on. Unfortunately one of them laughed loudly at some joke or other, with the result that their presence was discovered and they were invited to come inside, and that was the end of smoking concerts for men only. This reminds me of another story about the Selangor Club when it had moved into larger premises on the site of the present Club. An amateur performance was being given and a popular official, who later acted as the Governor of a Colony, occupied the stage. His role was a comic one. The audience laughed, the audience roared! Never had a performer met with such a reception! The house rocked with merriment! The performer thought 'by Jove, I am making a hit', until he found himself hauled back in to the wings to be told that his pants had split.

St. Andrew's Night (1894)

One thing we must chronicle, some guileless youth with a pigtail—who ought to have been fighting for his country at this critical juncture in her history, or, better still, have lost his life for her, ere he could have done this thing—got hold of the haggis and cut it up for sandwiches!

John Russell, *Selangor Journal*, Vol. 3, p. 344 (the encounter with the donkey) and p. 103 (the massacre of the haggis); J. H. M. Robson, *Records and Recollections (1889–1934)*, Kyle Palmer, Kuala Lumpur, 1934, pp. 9–10 (concerts).

82
Entertaining the Navy

JOHN RUSSELL

The Royal Navy had a number of small vessels on patrol in the Straits, to 'show the flag' and, if necessary, chase pirates. HMS Mercury *was larger than most, a cruiser of 3,730 tons and 13 guns; 'when new, the fastest warship afloat'. Her arrival at the mouth of the Klang River in June 1895 was a major event in the Kuala Lumpur social calendar. Parties of officers and ratings were brought up to Kuala Lumpur by train and accommodated in a vacant hospital ward. In addition to the entertainments here described, they took part in a shooting match, a cricket match, and several football matches against local teams, and there were lunches and dinners at the Residency and on board the ship.*

The opening paragraph refers to the arrival of members of the Malay royal dynasty, brought up from Kuala Langat in the government launch to join in the fun. The half-jocular but always complimentary reporting of local musical performances was a reflection of the fact that the reporter had to live in the same small local society as the performers.

RAJAS Kahar, Yusuf and several others, however, gladly availed themselves of the opportunity, and the *Esmeralda* returned with them to Kuala Klang, where the Rajas were most hospitably received by Lieut. Pearson on board the *Mercury*, and spent a great part of the afternoon in going completely round the vessel, in whose appointments they shewed the keenest interest. The torpedoes, engine room, quick-firing guns, the tanks for condensing water and the luxury of the Captain's cabin alike attracted their attention, and, at length, Raja Kahar paused, and turning round, remarked in Malay, 'This vessel was never built by human agency!' a remark which was duly translated for the benefit of the hosts....

The Cigarette Smoking Concert on the night of the 18th [June 1895], organised by Mr A. S. Baxendale, gave a very enjoyable evening to a large audience. Had it not been for the kindness of Mr Tearle, who gave a general invitation to the men to attend a Smoking Concert at his quarters, the room would have been inconveniently crowded. A number of the sailors, however, attended both. Towards the close of the printed programme ... some songs and step-dancing by members of the ship's company were introduced. Unfortunately, there was no one present who could accompany the dances—a great drawback.... Mrs Haines and Mrs Travers gave a very charming rendering of the duet 'Greetings'. Mr Dougal is as great a favourite as ever with a Kuala Lumpur audience, he sang 'Sweet Marie' with great expression and taste, and was awarded an encore. Mrs Haines is always a strong line on a concert programme and her singing of 'Whisper and I shall hear' was much appreciated by her audience, and 'My Dearest Heart' was given by her in answer to repeated applause....

On Wednesday morning the members of the Selangor Hunt Club arranged a meet for the entertainment of their friends from the *Mercury*, the notice said 'Corner of Maxwell Road, 6 a.m.'. Sharp at that hour a strong party mustered with guns and were soon posted by the Master, who proceeded to draw the jungle near Mr Paxon's garden for pig—unfortunately, they were not at home. A move was then made towards the Selangor Coffee Estate, where Mr Hampshire is living, and here the dogs put out two *kijang* [barking deer] which were bowled over in grand style, right and left, by Mr W. D. Scott, the shootist being pardonably proud of his excellent marksmanship. No time was lost in getting the dogs to work again, and very soon they were in full cry after pig, one attempted to cross the railway but was brought down by a clinking shot from Mr Youel RN, another was fired at and missed. There were any quantity of pig on foot, but owing to the very dry weather the dogs had great difficulty in following up their game. An unfortunate accident happened to one of the best dogs, his foot being terribly bitten by an enormous iguana; this brute was shot by the dog-boy and measured over five feet. By 11 a.m. our visitors were about tired of the sun, and a move was made to the 'Spotted Dog' for refreshments: the result of the morning's sport was sent on board the *Mercury* by midday train.

John Russell, 'HMS Mercury at the Kuala', *Selangor Journal*, Vol. 3, pp. 334–42.

83
Taking the Air

JOHN RUSSELL AND MRS STRATTON BROWN

The lavish use of domestic staff, as here described, was possible only so long as wages (typically Malayan $8–10 per month at this time) were low. By the 1930s wages had increased three- or fourfold and the household staff had fallen to a mere half dozen, including a driver and a gardener.

The Straits Settlements had been an Indian dependency until 1867 and Indian words, such as ayah *(for a nursemaid) had not yet been replaced by* amah *(a Chinese female domestic). The essential feature of local leave was getting up the hill to a refreshing drop in temperature. In a period when there were no motor cars, the hill station had to be within easy reach. Later on Fraser's Hill and the Cameron Highlands, at a greater distance but more spacious and offering better recreations, such as golf, displaced the local resorts of Treacher's Hill in Selangor and Larut Hill (near Taiping) in Perak and, of course, Penang Hill (see Passage 30 above).*

Perambulation

THE dusky nursemaid of the East, or of Kuala Lumpur, at any rate, is rather better off than her fairer sister of suburban London when taking her charge out for an airing. (We use the singular advisedly, because in the same way that a syce declines to look after two ponies, so does an ayah declare it impossible to attend to more than one child).... The ayah is provided with a companion in the shape of a young man, whose duty it is to wheel the carriage. Why this should be, it is hard to say.... We must suppose that the necessity arises for a young man to assist the nurse in taking one little child out for an hour in the evening, in the same way that the cook here finds it necessary to have a young man to carry home the purchases from the market, to light the fire, to prepare the vegetables, to—in fact, do almost all the work while he smokes—the necessity of doing as little real work as possible. It is the same all round. The house boy would faint if told to wash plates or fetch water, and one gets quite diffident about asking the 'tukang kabun' [gardener] to run with a message.

Up the Hill

For a big holiday Bukit Kutu (called also Treacher's Hill) was the choice. One went by rail to the terminus at Kuala Kubu, walked, rode or was carried seven miles to the top of the hill by jungle track to a nice bungalow.

Usually two families applied for leave. One for the first fortnight (the time permitted for use of the bungalow by Government) the other for the second, and so the women part of the families had a month with their children—the men folk getting a fortnight. Of course one had to take provisions.

On one of these expeditions on which I went, two sheep were driven up, crates of fowls, ducks, etc. carried by coolies, all tinned provisions in boxes. One took one's houseboy, and, if there were children, amahs or ayahs.

It needed about 20 coolies for the baggage, so it became a minor mountaineering proposition. Most of us walked the whole way but the older women and babies were carried in chairs slung on poles, four carriers to a chair.

From the station to the foot of the hill was fairly flat and rickshaws were taken for that, then the six miles up the track began. There was a fine durian orchard in the lower part belonging to the Sultan and ordinary durians could be obtained, but not a very special variety.

We left Kuala Lumpur by the early morning train, got to Kuala Kubu about 9 a.m., had breakfast at the rest house, and then went six miles up hill on jungle tracks, single file. It took a long time but we generally managed to get up by 1 p.m., just before rain was likely.

Then there was settling in all the baggage, poultry and other livestock. There were four big bedrooms and a dining room and living room, kitchen and servants' quarters. Fires were lighted for it was chilly.

It is between three and four thousand feet to the actual top, from which one gets a magnificent view.

We were told to bar doors at night as wild animals prowled round. Monkeys, etc. always made a lot of noise at dark settling on to their perches.

There was a tennis court marked out and fresh vegetable plots were looked after by the caretaker and his mate.

John Russell, *Selangor Journal*, Vol. 1, p. 35 (nursemaids); E. Stratton Brown, 'Looking Back on Selangor in the Nineties', *Fifty Years of Progress 1904–1954, Malay Mail Supplement*, 1955 (on Treacher's Hill).

Animals

84
Cuddling Up to a Tiger

ARTHUR KEYSER

Until the end of the nineteenth century tigers were a real menace in the rural areas of the Malay States. There were tiger stories of all kinds: horrific (Passage 86 below), magical, and, as in this passage (and Passage 57 above), tinged with humour. This anecdote is from Keyser's recollections of his time in the remote Jelebu district of Negri Sembilan. (See also Passages 86 and 88 below.)

YET some people go mad without achieving content, as the following story will show. A tall young Eurasian surveyor once served under me. In the performance of his duties he had to visit remote places in the jungle, remaining absent many days at a time. On one occasion this stay was so prolonged that his servants reported it to me. On search being made for him, he was met on the following morning, returning, as he explained, by easy stages. But his mental condition seemed queer. This was partially accounted for when the Malays with him stated that he had suffered a severe shock. And then they told me how he had been accustomed to sleep in a little 'atap' shelter on the ground, not raised on a platform, as was the custom, for he liked to lie near the embers of the fire, since it was often chilly after dark. One night, finding the fire unusually warm, he nestled more closely against the fragile wall. At daybreak there was a commotion and alarm, as the Malays screamed, 'Rimau, rimau!' (tiger, tiger). Seizing his gun the surveyor rushed out of the hut and fired hurriedly at a retreating form, just visible in the distance. Then they told him that the animal had been first noticed apparently enjoying the warmth, as he lay between the embers and the hut. A close inspection of the spot clearly showed this story to be true. From the moment that the poor surveyor realised that he had spent the night cuddling up against a tiger for protection from the cold, he became odd, 'and this, Tuan, is why you find him thus', concluded the narrator of that astounding tale. However, we led him home and for a time this night-tiger appeared to have left him. But not for long, since a few days later, on Christmas Day, a note was brought to me in the morning which ran, 'I have just drunk iodine to poison myself and wish you a very merry Christmas.' I felt something less than merry as I rushed for my horse and galloped for the surveyor's house to find that the first part of his message was literally true. In the hope of making him sick I took a feather and forcing his mouth open proceeded to tickle his throat. This produced no effect on the patient, but so reacted on me that I was obliged to seek the open window and be violently ill. However, my efforts and remedies combined succeeded in saving his life, and when the poor man seemed in a fair way towards recovery I was able to return home and resume the merry celebration of Christmas he had wished for me, but had so nearly succeeded in destroying.

A. L. Keyser, *People and Places: A Life in Five Continents*, John Murray, London, 1922, pp. 131–2.

85
A Cock-fight

HUGH CLIFFORD

As with other sporting contests (see Passage 78 above), cock-fighting flourished because of its intrinsic interest and also because of the excitement of betting on the result. In Clifford's description the fight is staged in the open space before the Sultan's istana, under the supervision of the royal pages ('the

King's youths') as attendants in the Sultan's household. Clifford had witnessed such events during his first two years (1887–9) as British 'Agent' (representative) at the court in Pekan.

A cock-fight between two well-known birds is a serious affair in Pahang. The rival qualities of the combatants have furnished food for endless discussion for weeks, or even months before, and every one of standing has visited and examined the cocks, and has made a book on the event. On the day fixed for the fight, a crowd collects before the palace, and some of the King's youths set up the cock-pit, which is a ring, about three feet in diameter, enclosed by canvas walls, supported on stakes driven into the ground. Presently the *Juara*, or cock-fighters, appear, each carrying his bird under his left arm. They enter the cock-pit, squat down, and begin pulling at, and shampooing the legs and wings of their birds, in the manner which Malays believe loosen the muscles, and get the reefs out of the cocks' limbs. Then the word is given to start the fight, and the birds, released, fly straight at one another, striking with their spurs, and sending feathers flying in all directions. This lasts for perhaps three minutes, when the cocks begin to lose their wind, and the fight is carried on as much with their beaks as with their spurs. Each bird tries to get its head under its opponent's wing, running forward to strike at the back of its antagonist's head, as soon as its own emerges from under its temporary shelter. This is varied by an occasional blow with the spurs, and the Malays herald each stroke with loud cries of approval. *Basah! Basah!* Thou hast wetted him! Thou hast drawn blood! *Ah itu dia!* That is it! That is a good one! *Ah sakit-lah itu!* Ah, that was a nasty one! And the birds are exhorted to make fresh efforts, amid occasional bursts of the shrill chorus of yells, called *sorak*, their backers cheering them on, and crying to them by name.

Presently time is called, the watch being a small section of cocoa-nut in which a hole has been bored, that is set floating on the surface of a jar of water, until it gradually becomes filled and sinks. At the word, each cock-fighter seizes his bird, drenches it with water, cleans out with a feather the phlegm which has collected in its throat, and shampoos its legs and body. Then, at the given word, the birds are again released, and they fly at one another with renewed energy. They lose their wind more speedily this time, and thereafter they pursue the tactics already described, until time is again called. When some ten rounds have been fought, and both the birds are beginning to show signs of distress, the interest of the contest reaches its height, for the fight is at an end if either bird raises its back feathers, in a peculiar manner, by which cocks declare themselves to be vanquished....

The victorious bird ... is carried off followed by an admiring, gesticulating, vociferous crowd, to be elaborately tended and nursed, as befits so gallant a bird. The beauty of the sport is that either bird can stop fighting at any moment. They are never forced to continue the conflict if once they have declared themselves defeated, and the only real element of cruelty is thus removed. The birds in fighting, follow the instinct which nature has implanted in them, and their marvellous courage and endurance surpass anything found in other animals, human or otherwise, with which I am acquainted. Most birds fight more or less; from the little fierce quail, to the sucking doves which ignorant Europeans, before their illusions have been dispelled by a sojourn in the East, are accustomed to regard as the emblems of peace and purity; but no bird, or beast, or fish, or human being fights so well, or takes such pleasure in the fierce joy of battle, as does a plucky, lanky, ugly, hard-bit old fighting-cock.

Hugh Clifford, *In Court and Kampong, being Tales and Sketches of Native Life in the Malay Peninsula*, 1st edn., 1897, 2nd edn. with autobiographical preface, Richards Press, London, 1927, pp. 48–51.

86
Trapping Tigers

DANIEL VAUGHAN

The tiger was perceived as a menace, both from the risk of death and as a source of magic, as here described. Against the tiger Malay villagers deployed both their technical skills in trapping and magical protection. (On Vaughan, see Passage 49 above.)

FIRE-ARMS are in common use and some Malays amuse themselves with bird shooting. They however usually prefer large game and many are exceedingly bold and expert in killing tigers. It is said that a tiger becomes extremely fierce and bloodthirsty when once he has tasted human blood and will seek his prey in the most crowded campongs and will watch a house night after night in the hope of catching a man. A remarkable instance fell under the writer's observation. Several men had been killed at a village in Province Wellesley by the same tiger and for several nights he had been heard prowling about the houses regardless of cattle and dogs that fell in his way. He was evidently bent on catching one of the inhabitants; finding at length that the villagers kept close, he actually sprang at the door of a house at night, burst it open, seized a man from his bed and walked off with him. At daylight he was traced by his foot-prints into the jungle and the body of the man was found partly devoured. A famous shot, one Eting, a samsam (or cross between a Malay and Siamese) was in the neighbourhood and he proposed that the remains of the poor fellow should be kept in the house as the tiger would be sure to return for a second meal. This was done and over the door of the house a strong platform was erected on which Eting took his station with his guns. Sure enough the tiger a little after nightfall returned to the house and was shot through the head. Tigers are frequently caught in traps—the most common is the pit trap which is used in all parts of India. A deep pit is dug and the bottom staked with sharp pointed staves. The mouth of the pit is concealed by branches and leaves and the bait (a dog generally) is tied to a bar over the centre. The tiger in prowling about discovers the bait, naturally springs at it and alights on the stakes, he is often pierced through by them—if not he is easily despatched with long spears.

Another trap employed in catching tigers, resembles the figure of four traps used by school boys. The trap is made with poles cut from the jungle; the part or lid that falls is laden with logs till rendered so heavy that the largest brute is unable to raise it. The lid is held up by an upright post so placed that the slightest push will remove it. To this upright post the bait is fixed, which the tiger seizes and in endeavouring to drag it away, he pulls the post aside and brings the lid down. To prevent the lid crushing the animal a cross bar is placed on posts a sufficient height off the ground to protect the brute but not leaving enough room to permit him to rise; his captors then introduce their hands into the trap and tie his legs together and to prevent him biting a piece of wood is lashed across his mouth. The lid is removed and the animal is so powerless that he may be easily removed in a basket or slung to a pole. In this fashion he is carried to town and disposed of. No fixed reward is paid by government for the tigers destroyed but the fortunate sportsmen are enabled to dispose of them for a handsome sum. Chinese are the usual purchasers; to them the claws, teeth, flesh and bones are invaluable. The two former are strung on threads and worn about the person, or treasured in their houses as charms, the bones are calcined and ground to a fine powder and used as medicine in various diseases and the flesh is eaten to render them brave and hardy. The skin usually falls to the lot of a European. It occasionally happens that the master of a merchantman purchases a tiger for the English market and then the hunters reap a golden harvest. A tiger is usually sold for five and twenty or thirty dollars....

They believe that a kris of a certain shape and make protects the inmates of a house from danger while within the house, another shape renders the wearer invincible. A few verses of the Koran packed in an amulet is a charm that renders the wearer invulnerable. A remarkable instance of this superstition occurred about two years ago. A tiger had strayed from the main jungle into the populous village of Tellok Ayer Tawar, Province Wellesley, and caused much consternation among the villagers, who turned out armed to the teeth to destroy the beast; the tiger was quite as much alarmed at finding himself so far from his lair and took refuge in a patch of mangrove and nipah jungle that lies between the village and the sea; a cordon of villagers was drawn along the road to prevent the animal's return to his fastness. For sometime not a man was found courageous enough to enter the mangrove. At length an old man, bent with age, came forward and offered to attack the beast unassisted; the bystanders tried to dissuade him; but he was firm, his kris was invincible and a charm that he had about his person protected him, he would go, and he did with the greatest composure, he found the tiger, walked up to him deliberately and stabbed him with his kris; in an instant the tiger struck him to the earth and in a few seconds the old man's belief would have been dispelled for ever, but assistance was at hand: three or four young fellows armed with muskets had followed close on the old man's heels and one of them shot the tiger and rescued the confident old Muslim, his neck was found to be fearfully lacerated but the catastrophe did not check the old man's faith in his kris and charm; he believed that they had killed the brute. He ever continued to boast of his prowess and infallibility of his weapons, quite forgetting the opportune aid of the young man with the musket. As a reward the old man was made Pangulu Mukim [headman] of the village he resided in by government, which served to strengthen his superstition rather than decrease it; the young man that saved his life was also rewarded with a douceur of a few dollars.

J. D. Vaughan, 'Notes on the Malays of Pinang and Province Wellesley', *Journal of the Indian Archipelago*, New Series, Vol. 2, 1857, pp. 141–2 and 164–5.

87
Catching a Crocodile

WILLIAM HORNADAY

A Malay village typically stood on the banks of a river, which in addition to providing communication with other settlements was a water supply, a fishery, and a place for bathing and washing clothes. Hence, the villagers were often at risk of attack, usually fatal, by crocodiles, which lurked in large numbers in and near the rivers.

Hornaday (see Passage 45) had come to Jeram mainly to collect crocodiles (which he reduced to skin and skeleton after trapping them) for American museums. For him, the standard European method of killing crocodiles by shooting them from a distance did not suffice, since the wounded beast usually sank in the stream. Hornaday needed to trap crococdiles, and—as he had done in India earlier in his tour—he made use of local native methods of trapping.

ACTING under the advice of a Chinese fisherman who seemed to know how to catch crocodiles with a hook and line, we got a rattan about forty feet long for a line, and a dry cocoanut to tie at one end as a float. The Chinaman then proceeded to make an 'alir', such as the Malays use in Sarawak, by whittling an inch piece of tough green wood ten inches long into a shape something like a crescent, sharp at both ends and with a groove running round the stick at the middle, which was the thickest part, where the line was to be attached.

Some soft but very tough green bark was then procured from the jungle, and braided into a line six feet long, which was at one end fastened firmly round the middle of the alir, and at the other to the long rattan rope. This bark line was supposed to be so soft and tough no crocodile could bite it in two. The bait used was the body of a sting ray caught by one of the fishermen, which was lashed securely to the alir, one end of which was then bent up close to the bark line and tied to it with a bit of string that could be broken by a slight pull. The intention was that the alir should be swallowed point foremost, and when we pulled on the line the upper point would catch in the middle of the stomach, break the string and instantly bring the alir crosswise in the crocodile's interior.

The crocodile we wanted to catch was well known by his repeated appearance at the village, within a stone's throw of the houses, and he was described as being a perfect monster, with a throat large enough to swallow a large-sized man instantly. The villagers manifested great interest in our effort, and helped us in every possible way.

We took our tackle just far enough above the village to be out of sight, for we wanted our victim to have so good an opportunity that he would not feel bashful. Following the custom of the Malays we found an overhanging branch, quite low down, over the end of which we threw our line so that the bait hung within six inches of the water at high tide, and so adjusted that a very slight pull would bring it down. The rattan line we threw into the stream with the cocoanut buoy at the end, and quietly retired to the village to await developments.

At the close of the day the bait still hung there undisturbed, and I walked home to Jerom hoping for better luck on the morrow. The next morning we were there soon after sunrise, and the Chinaman joyfully informed us that the bait was gone. We got into a small Malay sampan and paddled up the creek at once to investigate. We found the cocoanut moving slowly through the water against the current and upon laying hold of the line we felt there was big game at the other end. We gave a vigorous pull, and the next instant were almost capsized in mid-stream by a pull we got in return. We then passed the line over the stern of the canoe and while I held it, the rest began to paddle down stream toward the village where we proposed to land our catch.

Then he showed himself. He rose to the surface apparently to see what was the matter, and, after giving a good look at us, started forward and began to turn as if about to go up stream. Before he had turned half round he fetched up with a violent jerk which must have given one point of the alir a vicious dig into the side of his stomach; for he began to plunge and thrash around with great violence, sending the water circling around him in huge waves. There was also considerable excitement at our end of the line, for the sampan was small, light, very tipsy, and contained three men of good weight. Chinaman, Malay, and Anglo-Saxon, each shouted at the other two in his own language. Had we been capsized I scarcely know which would have disgusted me most, the ducking in that dirty creek, full of crocodiles, or the loss of my rifle. As soon as we could I tied the weapon fast to the boat so that in the event of a mishap I would not lose it.

After this struggle the crocodile seemed to give up the fight, for he allowed himself to be towed down to the village without further resistance. But as we neared the landing place where we intended to haul him out, he made a final and still more vigorous struggle to get free. He snapped his jaws together in an effort to cut the line, but it was no use, so shutting them together like a vice he plunged first to one side and then to the other, striking out with tail and legs, diving deeply one moment and suddenly thrusting his ugly snout far out of water the next.

Another boat came to our assistance at this point and the huge old reptile was dragged shoreward by main force. The men landed and dragged him close up to the shore without further resistance on his part, whereupon I fired a bullet into his neck from the side which cut his spinal marrow so neatly

One more for the museum. Hornaday lands his crocodile, from William T. Hornaday, *Two Years in the Jungle: The Experiences of a Hunter and Naturalist in India, Ceylon, the Malay Peninsula and Borneo*, Charles Scribner's Sons, New York, 1885.

that the vertebra was but very slightly injured. He was the very crocodile we wanted, and his death occasioned no sorrow. He measured exactly twelve feet in length, and his weight was four hundred and fifteen pounds. He was so old, so dingy, dirty, and ugly every way that I concluded to take his skeleton instead of his skin, and spent a day in roughing it out neatly.

William T. Hornaday, *Two Years in the Jungle: The Experiences of a Hunter and Naturalist in India, Ceylon, the Malay Peninsula and Borneo*, Charles Scribner's Sons, New York, 1885, pp. 305–7.

88
Tiger against Buffalo

FRANK SWETTENHAM

In addition to cock-fighting (see Passage 85 above), duels between buffaloes were a common amusement in the north-east of Malaya. Skeat (see Passages 40 and 41 above) witnessed a 'bull-fight' in Kelantan, noting that the spectators became wildly excited though the fight was a rather tame affair in which neither animal suffered serious injury. Setting one kind of animal to fight another as a spectator sport was less common, could be no less of an anti-climax, and offended European tastes.

Swettenham here relates an episode early in his Malayan career when he was sent to escort, and act as interpreter, during the visit to Johore Bahru of a Russian Grand Duke. Maharaja (later Sultan) Abu Bakar of Johore always did his best to entertain European visitors of high rank (see Passage 46 above) but on this occasion his efforts misfired somewhat.

IT was not long after I had passed in Malay that a Russian Grand Duke visited Singapore, and for his entertainment the Maharaja of Johore gave a great luncheon at his Johore residence, and staged a buffalo and tiger fight to follow. I was one of the large party of guests from Singapore and, being of no account, was lunching in a side room when, without previous notice, I was summoned to the banqueting hall and told to stand by the Maharaja and interpret the Malay speech he was then going to deliver to welcome His Imperial Highness and propose his health. As soon as I reached his chair the Maharaja stood up and began to speak. He was not accustomed to interpretation, and did not think it necessary to pause in the flow of his eloquence, so I had to ask him to wait whilst I put into English long passages, not easy to follow. His Highness soon jumped to the situation, and an ordeal rather trying for me was concluded with, what appeared to be, great satisfaction. It was an ordeal because I had no warning of what was expected of me, and because the room was full of Malay officials and of English residents from Singapore who had lived there for years and regarded the 'Cadets' as a class of Government servant not to be taken too seriously. I was quite aware of the criticism to which I was being subjected, but, fortunately, my memory enabled me to remember the Maharaja's long sentences, and, as I had no difficulty with the language, the incident ended happily.

The show which had attracted to Johore this large gathering of Malays, Europeans, Chinese, Indians, Arabs and other nationalities was a rather sorry affair, and so far as I know, was never repeated. A plot of ground, oval in form, enclosed by a fencing of heavy palisades, with enough space between each to enable spectators outside to see what was going on within, formed the arena. A portion of this gigantic cage was curtained off by canvas, and into that compartment was carried a caged tiger which had been caught recently in a pit-trap.

A multitude of spectators surged round the enclosure, and, when all was ready, an ordinary water buffalo, taken from the shafts of a cart, was introduced into the unoccupied section of the ring. The buffalo walked quietly to the centre of the enclosure, and stood there while the curtain was removed and the tiger disclosed to his opponent's view. The two beasts stared at each other, and the hairs of the buffalo's neck began to stand up but, almost immediately, the tiger started to run round the arena, keeping close to the fence. The movement evidently determined the buffalo to attack, and, judging distance and his enemy's pace with accuracy, he rushed head down at the tiger, transfixed him with his horns and smashed his body against the palisade. Then, stalking backwards with measured strides, he withdrew to his position in the middle of the arena, all the while keeping his eye on the tiger which lay, grievously damaged, against the palisade. After a few minutes the tiger got on his feet and started again to run, as though he hoped to find a means of escape. He had only travelled half the circumference of the enclosure when the buffalo charged him again, impaled him with its horns, crushed his body against the palisade, and finally threw his now helpless victim in the air, caught the falling body on his horns, and threw it violently on the ground. I did not want to see any more; but my impression at the time was that certain spectators near the combatants prodded the stricken beast with sticks and umbrellas in an attempt to get him running again. He was, however, past that or any other effort, and did not move again. I was curious to see what damage the buffalo had sustained, and, beyond an unimportant scratch on his nose, there was nothing. I ascribe that rather astonishing result to the fact that the tiger's home was in the jungle, far from men and their haunts, whereas the buffalo was a tame beast accustomed to human beings. Therefore, when stripes found himself in a pen, surrounded by moving and shouting people, he disregarded the buffalo and sought only to get away. The buffalo, being at home and at his ease, recognized in the tiger a dangerous enemy, and proceeded to deal with him. Years later in Pahang, a Malay told me how he watched a tiger stalk a herd of semi-wild buffalo feeding in an open space of grass surrounded by forest. The herd got wind of the tiger and a great bull stalked out alone and faced the enemy while the cows and calves stood grouped together to watch the proceedings. When the bull was near enough to charge, the tiger sprang on his shoulders and tried to seize the great beast by the neck; but the bull threw him off, charged him with head down, and the tiger, having had enough, made off to the jungle.

Frank Swettenham, *Footprints in Malaya*, Hutchinson, London, 1942, pp. 17–18.

Wars and Troubled Times

89
Guy Fawkes—Malay Style

MUNSHI ABDULLAH

In the nineteenth century, the Malay States fought wars with skill and determination but, as in other times and places, there were conventions to be observed. Communications and logistics imposed limits on the numbers of men who could be deployed and on the material which could be used. There was much manoeuvre and casualties were generally not heavy.

The outbreak of civil war in Kelantan in 1838 led to the detention there of junks loaded with goods belonging to Singapore merchants, and once again (see Passage 13 above) a mission was sent to negotiate with the Ruler. We find Munshi Abdullah (see Passage 62) at a later stage in his career; he was now a prominent figure in the Singapore Malay community. The merchants employed him to deliver the Governor's letters to the leaders of the warring factions. Thus, Abdullah came to write the second of his major works—an account of his voyage up the East Coast in 1838. On his arrival in Kelantan, Abdullah found that the fighting had come to a deadlock in the form of a siege of a stockade, not an uncommon event in Malay warfare. One side sheltered behind its fortifications and the other preferred to starve them out rather than risk many deaths in an assault. The besiegers sought Abdullah's advice on how to resolve the impasse.

WHILE I was sitting in the cutter, a Haji came to see me; he was a Trengganu man and bore the title of Panglima Besar, or Warrior Chief. After some casual conversation, he said that the Raja Bendahara had told him to ask me whether I knew the ingredients of gunpowder.

I told him that I did not know. I added that I had a book full of scientific information, such as how to make gunpowder, how to make eye-lotion, how to plate metal, how to inlay krisses and how to make loaf-sugar; but the book was not with me—I had left it at Malacca.

He said that he did make gunpowder once, but it didn't turn out well.

'I have a very difficult problem,' he said. 'And the Raja too can't think of any plan to smash or break down the enemy's fortification; he can't stand this continual cannonade, and furthermore the enemy are in a position to see everything that we do.'

'This isn't war,' I replied, 'it's playing at war. And it is the poor and the peasants and the foreigners who suffer for it; this business would be finished in a day if it were a European war. Men who go to war should not be afraid of death and dig holes in the ground to live in. They fire four or five shots, and then both sides stop to eat and drink; and then they fire again. By their methods no decision can be reached—no, not in ten years; each side sits in its stockade, and even if there were ten koyans of gunpowder, it would [not] be finished. Why don't both sides go out to an open space with their cannon and rifles and face up to one another? Then it wouldn't take long! The lucky man would win, and the war would be over!'

'There was a time,' he said, 'when the enemy sent us a challenge to a hand-to-hand battle, but our people wouldn't consent. Later the Raja Bendahara sent several similar challenges to them, but they wouldn't consent. And so nothing came of it. Furthermore the Raja Bendahara said that he did not wish to harm his subjects, and that kind of warfare would harm many of them, and so he preferred to besiege the enemy and prevent food from reaching them; in the long run they were bound to give it up and come out.'

'In that stockade,' continued the Haji, 'you couldn't get a betel-leaf or a fish if you offered a dollar; they are eating rice-refuse. And I will tell you that every night deserters come from the stockade, ten or fifteen men a night. They all come to the Raja Bendahara and he does nothing to them.'

I asked the Haji how many men he thought there were in the besieged stockade, and he said that there were five or six hundred spearmen and the rest were Chinese, but the amount of property in the stockade was very large; the other three Rajas according to him had tens of thousands of men.

I asked him which Raja the people preferred, and he replied that everyone preferred the Raja Bendahara.

But he said that there could be no final settlement until the return of the missions to Siam; the Raja of Siam would decide who was to be the Raja of Kelantan. All four Rajas, he said, had sent missions to Siam with five or six thousand dollars as a present to the Raja there.

Then he asked what plan I could suggest for demolishing the enemy fortification.

I said that I knew a way in which the whole structure and the men in it could be destroyed in one minute.

He asked me to explain my plan.

In reply I asked the distance between our stockade and the enemy's, and he said, twenty or thirty yards. I said that, if so, it was simple. He said that he would be greatly obliged if I would tell him what should be done, so I explained.

'Dig a hole inside our stockade,' I said, 'and thence make a tunnel until you are right under the other Raja's house. Then bring four or five barrels of gunpowder, put them in position, and join them with a fuse passing through a hole in each. Ram the earth well down on them, taking the end of the fuse out with you. Then, when they are all gathered together for food or whatnot, light the fuse. In an instant they will all fly skywards.'

'You know how strong and thick and broad the walls of the fort at Malacca were; every stone was granite and six feet across! Well, Major Farquahar sent it all flying in the way I have mentioned; I saw him do it. If instead of doing it in that way, he had taken it down stone by stone, the job wouldn't be finished yet.'

The Haji was delighted and said that he would convey my suggestion to the Raja. He went off, and during his absence the men who had been ashore to buy provisions came back. Then we made both cutters ready for departure.

Then the Haji came back and said he had put my suggestion to the Raja. The Raja, he said, was truly delighted with the idea, but averred that he did not like to carry it out, as it would cause the deaths of so many of his own people; if only the other Raja were killed, it wouldn't matter.

The Haji added that he could not see how the war could be finished quickly. He pointed out that the men in the two opposing stockades were not strangers; in some cases the wife was in one stockade, the husband in the other; the father in one, the mother in the other; the elder brother in one, the younger brother in the other. 'They can all see one another,' he said, 'so how can they bring themselves to kill? But they do what the Raja tells them from fear of him!'

'All right,' I said, 'If so, let the Raja go out and challenge the other to meet him in single combat with gun or kris. Then it would soon be settled, and this waste of money on ammunition would stop.'

'Since the beginning of the war,' said the Haji, 'you have no idea how many koyans of rice have been supplied by the Raja to his chieftains and personal guard! Three chests of opium too! And all to no purpose!

I haven't forgotten your idea; it's worth a thousand dollars. But it is awkward for us, being all related.'

Abdullah bin Abdul Kadir, Munshi, *Kesah Pelayaran Abdullah*, translated by A. E. Coope, *The Voyage of Abdullah: A Translation from the Malay,* Malaya Publishing House, Singapore, 1949, pp. 46–8.

90
The Siege of Klang in 1870

WAN MOHAMED AMIN BIN SAID AND PENGHULU HAMZAH

The siege of Klang, at the beginning of the Selangor civil war, was one of the largest and most protracted episodes among the many conflicts in the Malay States in the nineteenth century. It was also marked by the large-scale use of artillery for the first time. To explain the two narratives which follow, it suffices to say that Raja Mahdi had seized the town of Klang, driving out the former chief of the Klang district. But the latter's sons sought the support of Tunku Kudin of Kedah, who was the son-in-law of the Sultan of Selangor and seeking to govern the State in the Sultan's name. This coalition besieged Raja Mahdi and eventually forced him to flee from Klang.

The father of Wan Mohamed Amin was a follower of Raja Mahdi, and so the two—father and son—were at first in the town, though they later decamped to join Tunku Kudin and the besiegers. Kudin brought down levies from Kedah, to strengthen his forces, and bought munitions with money borrowed from Chinese backers in Malacca. Penghulu Hamzah was one of Kudin's subordinate commanders.

Wan Mohamed Amin's Story

I went three times with the fighting men. The first time was in the fighting behind Bukit Jawa, i.e. behind the DO's [District Officer] house. One day the defenders obtained information that the enemy had built a stockade right on the 5 way junction on the Langat road. About 2 p.m. the defenders went out to attack it and I went with them, with a cousin's brother-in-law called Raja Ngah Malim who was one of the leaders.

The two sides fought in the lalang [coarse grass], undergrowth and young trees as thick as one's arm, in brief there was nowhere where one could be protected. So, in positions of danger one could not walk erect but had to crawl. The bullets flew like rain—I saw a man hit by a bullet because he was walking upright, not wishing to crawl. His body was carried away in a sarong back to the fort. After the fighting had lasted three quarters of an hour the enemy retreated and their newly built stockade was taken; it had one side only.

The second time I went with the fighting men was to the source of the Telok Pulai river. There the enemy had built a stockade of banana trunks because there were no trees. The defenders heard report of this about 4 p.m. and they went down to attack it. I went in company with Panglima Raja Ngah Malim. From what I saw I would say that the fighting was not serious because, although the distance between the two sides was only 60 to 90 feet, there were many coconut trees behind which each man took shelter, one man one tree. Such was the fighting that day.

The fighting lasted about half an hour. A young Bugis called Salleh was hit in the head by a bullet. His body was carried back to the fort and I went with the carriers. Another young man, Sidin, was hit by a bullet but he did not die. He too was carried back to the fort. The enemy could not hold out in their stockade of banana stems because, when the bullets hit it, they penetrated right through. Not long afterwards parties of the enemy came up from Sungai Pinang and Bukit Ellen. By then it was almost dark and all the men from the fort withdrew into it.

The third time I went with the fighting men was when they attacked the stockade of the Mandiling men. By that time I had changed over to the side of Tunku Kudin. The Mandiling men had built a stockade at Tanjong Belit, on the downstream side of Damansara. They had made two stockades, one by the river bank and one under a fig tree. On the fig tree they made a hut and put in it a small swivel gun. In addition they had a stockade on the landward side. Tunku Kudin's men had a stockade at the jetty at Damansara under the charge of Dato' Mangku, a Rembau man. At that time Raja Ismail and his younger brother, Raja Empih, were at Damansara. They wanted to return down river but could not do so because of the enemy stockades. So Tunku Kudin ordered his fighting men to advance upstream to attack the enemy stockade. A barge was made ready with a gunshield of mahang wood, fitted like a canopy to resist the bullets. There was a small boat at the bow of the barge. The man in charge of the boat was a Malacca trader named Baba Ot. My father and I went along as followers of Raja Hassan.

When the barge and the boat came opposite the enemy stockade, there was a rain of bullets like the sound of frying rice. But we did not return fire but instead the boat and the barge went on up river to the jetty at Damansara. When the supplies for the men in the stockade had been unloaded, the boat and the barge returned downstream. When we were opposite the enemy stockade, the barge was anchored and the boat was ordered to go downstream. The gun and the muskets fired on the stockade. The men who fired were Pahang men under Encik Ambo. The fighting went on into the night. At about 8 a.m. the stockade was broken and the enemy fled from it. I do not know how many Mandiling men died but there was a lot of blood left behind. Not one man was killed or wounded on our side.

Penghulu Hamzah's Story

On his arrival at Alor Star [Tunku Kudin] summoned the best fighters of the day ... and made them his Panglima Besar (Generals) and ordered them to recruit men.... Several thousand men rallied round them, and when a sufficient force had been collected the expedition sailed in 72 prahus and junks, taking only a few lelahs (swivels) and small arms, as they were depending on Baba Tee Yee for arms and ammunition. On arrival in Klang a large force of local men joined them and the war started in earnest under the command of Tunku Kudin Kechil....

[Raja Mahdi] had fighting rafts built of logs of hard timber and made bullet and cannon-ball proof by bulwarks of mahang wood three layers thick, besides a schooner protected by shields made of logs of wood behind which the cannons were mounted and the men fought. This schooner and the rafts kept moving up and down river with the tides, and our stockades were continually under severe fire, especially when the rafts were made fast to the bank of the river and raked us at close range with such fatal result that we had to dig ourselves in, in the trenches. The keramat (charm) of the enemy was so potent that many a time our cannon balls went wide of the mark, although fired almost point blank at the rafts. The enemy could not storm our stockades because of the 'ranjaus' (sharp stakes driven into the ground to pierce the feet) made of bamboo and as sharp as razors. We were well armed and had plenty of ammunition. The stockades were protected by logs of wood stacked round them in pigsty fashion 18 feet in width....

There were cannons using balls 3 spans in circumference. They were loaded with a charge of black gunpowder, rammed down by coconut husk over which came the ball and then more coconut husk. The charge was fired by a live brand held to the loose powder in the touchhole. The report was deafening and terrific but after a time one lost all sense of fear....

I was in charge of the commissariat and also had to superintend the cooking of the opium, as a good many of our fighters were opium smokers. I was also detailed to mount the 'langkayan' (tower) to see with my glasses whether we hit the target or not....

[When Klang surrendered] the scene that met the eye baffled description. Men were howling and wailing and women weeping and shrieking. It was the despairing wail of human beings who had staked their all on the fortunes of war and were now being turned adrift into the world, without home or any earthly possessions. They were not molested but allowed to carry what few things they had. There was no looting. The Kedah Malays had strict injunctions to take no spoils.

Wan Mohamed Amin bin Wan Mohamed Said, *Pesaka Selangor*, edited by Abdul Samad Ahmad and translated by J. M. Gullick, Dewan Bahasa dan Pustaka, Kuala Lumpur, 1966, p. 47; J. C. Pasqual, article in the *Singapore Sunday Times*, 14 October 1934, telling the story of Penghulu Hamzah.

91
A Visit to Perlis in 1838

SHERARD OSBORN

As a result of the Anglo-Siamese treaty of 1826, Britain was reluctantly obliged to lend support to the Siamese forces sent to crush a Malay rising in Kedah in 1838. The support took the form of a naval blockade of the Kedah coast, to interrupt the flow of supplies to the Kedah Malays. The blockade was enforced by HMS Hyacinth, *acting as supply ship to a flotilla of smaller craft, one of them a gunboat commanded by the sixteen-year-old midshipman, Sherard Osborn.*

The passage describes his adventurous passage up the Perlis River, to collect fresh water. Haji Long, to whom he refers, and other Kedah Malay leaders, resented the British blockade but did not wish to make an open break with the British navy, since they drew their supplies from Penang.

ONE of the other gun-boats was despatched to seek water elsewhere, and I was ordered to start next morning in a large sampan, with a couple of empty casks, to procure fresh water above the reach of the tide in Parlis river. My perfect confidence in the Malays, in spite of Mahomet Alee's threats, enabled me to look forward to my cruise into the very heart of their territory without any feeling but that of great curiosity, and a pardonable degree of pride at being the first to see all the war-prahus.

Early in the forenoon I started, in a good sampan, with one English sailor, an interpreter, and six picked Malays, all well armed; but their muskets and pistols were placed where they would be ready for use without attracting attention. The flood-tide ran strong, and we swept with it rapidly up the stream; the first mile or two was very monotonous, the banks being for the most part mangrove, and another tree which seems to delight in an equally amphibious life. At a curve in the river we came suddenly on a stockade, and, being hailed immediately by some men on guard, I felt to what a thorough test we were going to put Malay chivalry.

The stockade across the stream was well and neatly constructed, having a couple of tidal booms fitted in such a way that the guard could at any moment, during either flood or ebb tide, stop up the only passage; and on either hand, some hundred yards back from the river, rose conical-shaped hills, on whose summits formidable batteries, constructed of heavy timber, commanded the stream in every direction.

The pangleman [panglima], or officer, at the guard-house smiled when I told him I was going up the river for water, and said he had no objection to my proceeding to Parlis to ask for permission; but, as to obtaining it, he laughed, and said all would depend upon the humour I should find Datoo Mahomet Alee in. Another three miles of fine open forest replete with Oriental interest now occurred, and the country improved in appearance after we had passed a spur of picturesque hills, through which the river had forced its way. Our approach to Parlis town was proclaimed by a line of war-prahus moored to either bank. The rapidity of the current, as well as my anxiety to reach the fresh-water point of the river, gave us but a flying glance at this much-talked-of and long-wished-for pirate fleet; and, besides which, I felt it desirable not to appear as if on a reconnoitring expedition.

They were handsome-looking craft, not very numerous, but with fine long guns mounted in their bows: they had but few men in each of them, though otherwise ready for sea.

Of Parlis we could not see much beyond that it was situated upon a plain on the south side of the river, and appeared capable of containing four or five thousand inhabitants. We pulled in for a light wharf constructed of bamboos, whereon an armed Malay had hailed to know what we wanted; and he, in reply to my answer that we wished to see the Datoo, said that was his house. I landed with two or three men, and, surrounded by a crowd of armed Malays, who hastened from all sides, was escorted to Haggi Loung [Haji Long].

That worthy received me, and said that Datoo Mahomet Alee was absent with his men fighting the Siamese: but what might be my errand?

I told him I was sent by my senior officer for water.

The Haggi laughed heartily for so holy a man, ... [and] with good humour told me to go: he would not stop me, but warned me to be careful, as all the country was in arms, and neither he or Datoo Mahomet Alee could be responsible for our safety.

That was all I wanted: so I bowed, and started back to the boat. Numbers of armed Malays—some of them, from their beautiful creeses [krises] and spears, doubtless men of importance—thronged the Haggi's anteroom and the pier; a few of them scowled in an unfriendly manner, and some of the younger game-cocks ruffled up, as if anxious to throw a feather with my men. I kept an eye upon them, however, and got all safe down, without any further interchange of civilities than a short address, which my English bodyguard made them off the end of the pier....

Unwilling to be detained, I pushed on as hurriedly as possible; and when we had gone, by my calculation, a distance of sixteen miles from the entrance to the river, another town, called 'Kangah' [Kangar], hove in sight....

A mile or so above the town, we arrived opposite some powder-mills, where a Malay sentry hailed us, and having told him we had Haggi Loung's permission to go for water, he did not detain us.

This fellow's confidence in his chief amused me. I asked him if Datoo Mahomet Alee was at Kangah.

'No,' he replied, 'he is on his march to Quedah [Kedah]!'

'How about the Siamese?' my interpreter asked.

'Pish!' said the sentinel; 'the Siamese! they will all be destroyed!'

We did not wait for further information, and, shortly afterwards, finding the water perfectly fresh, we being then about eighteen miles from the sea, we laid on our paddles, and filled our casks, bathed, washed, and drank water, with all the *abandon* of men who had long been strangers to the luxury of fresh water in large quantities. . . .

Our casks filled, we turned our head down the stream and dropped down to Kangah, where I purposed having our noon-day meal, and waiting for the tide to have ebbed sufficiently to ensure us a rapid passage down to the gun-boats. At a point just above the town, where some lofty trees threw a pleasant shade half across the stream, all the female population of Kangah, as well as the children, were enjoying a bath. . . . Choosing a convenient part of the river bank opposite 'Kangah', we made our sampan fast, and proceeded to cook rice for lunch. . . .

Whilst the rice was cooking, I thought I might as well run up and see the town: a boy volunteered to show Jamboo and me the bazaar and Datoo Mahomet Alee's elephants, and we accordingly started with a couple of followers.

The bazaar consisted of one narrow street, running at right angles to the river. Each shop had a sloping and open front, well shielded from the heat of the sun, on which was displayed the thousand strong-smelling fruits and vegetables, the gaudy Manchester prints, glaring red and yellow handkerchiefs, pretty mats and neat kagangs, piles of rice and tubs of ghee, handsome creeses, and formidable swords or choppers, which may be seen in all the bazaars of Singapore, Malacca, or Penang. There were Mahometan natives of the Madras Presidency, swathed in turbans and robes of calico—the embodied forms of the Great Moguls which figure on our playing-cards; greasy, black, and very strong-smelling Klings chattered, lied, and cheated as Klings only can do; Malays swaggered about, decked out in gay attire, and sporting beautiful arms and silver-mounted spears, looking so saucy and bold, that one felt half inclined to pat them on the back, and say, 'Well done!' for they knew as well as we did that their hour had struck, and all the scene would soon be dissipated like a dream, and they be pirating elsewhere.... No one could have supposed, from the scene in the bazaar, that fifteen thousand Siamese were close at hand, ready to impale, disembowel, or play any of the many pranks I have elsewhere related, upon each or all of those before me....

I now proposed to go to the elephants, which, from our guide's description, were at the other end of the town. We had just disengaged ourselves from the crowd, heat, and strong smells of the bazaar, when a general commotion occurred in the town, which had hitherto exhibited no signs of life except in the bazaar. Boys ran along screaming, women ran out in the balconies, and appeared very excited; and soon afterwards a large body of Malays, armed to the teeth, covered with dust, and looking much wayworn, passed rapidly along, marching, however, without order or military array.

I was informed through Jamboo, that it was impossible for me to visit the royal stables to-day, as some important event had evidently just taken place, and a great chief—possibly the redoubted Datoo himself—had arrived. I did not much care about pushing the point, as I was on shore on my own responsibility; and Haggi Loung's warning left me no excuse but that of curiosity, if we got into a scrape. One of my men sidled up to me, and said some of the natives were getting up a report that I was a spy, and that one of them had threatened him. I decided to return to my boat; and, from expressions which were uttered by those around, found it was high time I did so. Indeed I am not sure we should have escaped without a scuffle, had not a venerable-looking man joined us, and, by his authority, enforced a little more respect from the rabble....

The ebb-tide was running strong as I jumped into my boat, and casting off from the shore, we were soon 'spinning'—to use a seaman's phrase—down the stream.

Sherard Osborn, *Quedah: or Stray Leaves from a Journal in Malayan Waters,* Longman, Brown, Green, Longmans, & Roberts, London, 1857, pp. 260–79.

92
Square Bashing in Malaya

ARTHUR HUBBACK AND ROBERT BRUCE LOCKHART

During seventy years of colonial rule, various attempts were made to raise local forces for the defence of Malaya—not always with satisfactory results. But patriotic feeling among the European community at the time of the Boer War (1899–1902) led to the start of the FMS Volunteer Force. It thrived and other communities gave it their support, so that by the time of the Japanese invasion it was a useful territorial militia, whose potential was not recognized nor adequately used by the regular army commanders (see Passage 94). In 1933 a belated decision was taken to form a regular Malay Regiment, and its battalions fought courageously in the defence of Singapore.

Military efficiency in time of war is the product of sustained and well-organized training in time of peace. The passages which follow pursue that theme.

The Volunteers in 1905–6

THE enactment had provided in the annual efficiency qualification for attendance at a very large number of drills, and it was found that the sterotyped barrack-square parades were not well attended, owing to their uninteresting nature.... In April 1905 a scheme of reorganisation was laid before the Government. Briefly this provided for the erection of a semi-permanent camp in the vicinity of Kuala Lumpor, the holding of monthly camps of instruction in field manoeuvres, a reduction in the number of 'barrack-square' drills, and a special allowance of ball ammunition for musketry practice....

The monthly camps of exercise are held at the Volunteer Camp on the hills in the direction of Ampang, some three miles out of Kuala Lumpor, and adjoining the rifle range. The camp opens on Saturday afternoon, when some special scheme of field manoeuvres is carried out. Sunday morning is spent in squad and company drill and in musketry instruction. These camps are well attended. Special arrangements are made for transport, and all out-station members are allowed free railway passes to and from Kuala Lumpor. By reducing the uninteresting 'barrack-square' drills and instructing men specially in attack practice, outpost duty, skirmishing and scouting, it has been possible to enable each individual to understand how to take advantage of cover, to use his own initiative, and to realise the practical part of the instructions laid down in the infantry training. A squad of signallers has proved itself most efficient and exceedingly useful in all manoeuvres.

The Malay Regiment on Parade in 1935

Before us were a hundred and fifty recruits dressed in singlet and white shorts and the two hundred and fifty trained men, divided into three companies and looking very smart in their pale-green uniforms, with dark-green cap, pale-green web belt, shorts, puttees with Islam green tops, and well-polished black boots. From the laterite soil of the parade ground came a thin cloud of red dust. The air was charged with thunder. The heat was sweltering, and, as my head was splitting, I was prepared to be bored.

When the display started, I was pleasantly surprised. I am no soldier, but during my life abroad I have seen the ceremonial parades of most of the crack regiments of Europe. These Malay boys were marvellously efficient. They had the 'swagger' of guardsmen. Drilled by a Scottish Sergeant-Major, a short, red-faced Cameron Highlander with a hoarse whisper of a voice which sounded just as if he had completed a three-months' lecture in the United States, they performed a series of complicated evolutions not only with machine-like precision, but also with exultant pride. At the end they formed up with fixed bayonets about forty yards away and directly opposite to us. An officer blew his whistle, and suddenly the men charged with a chorus of blood-curdling yells.

I was quite sober, but as I saw this line of brown faces with open mouths, dilated nostrils, and fiercely gleaming eyes descending on us I was not quite sure what was going to happen. I felt a sudden uneasiness which was increased when Harry Rosslyn gripped my arm. Then the whistle blew again, and the men pulled up a yard or two away from us. It was a magnificent and even terrifying spectacle.

93
The Japanese Assault on Kota Bharu

MASANOBU TSUJI

This passage is taken from a book which first appeared in English translation in 1960. It is an account by one of the Japanese staff officers who planned and helped execute the Japanese assault on Malaya in 1941–2, leading to the fall of Singapore. The episode here described is the first Japanese landing on the beaches around Kota Bharu in Kelantan, which gave them their foothold from which to make further rapid advances.

AT midnight on 7th–8th December the three transports had anchored off shore, and despite the heavy seas that were running successfully transhipped their troops into boats. Then the naval escort began a bombardment of the coast as a signal to commence the landing. The enemy pillboxes, which were well prepared, reacted violently with such heavy fire that our men lying on the beach, half in and half out of the water, could not raise their heads.

Before long enemy planes in formations of two or three began to attack our transports, which soon became enveloped in flame and smoke from the bursting bombs and from shells fired by the shore batteries. The *Awagisan Maru* after ten direct hits caught fire; later the *Ayatosan Maru* did likewise after six hits. The officers and men of the anti-aircraft detachment, although scorched by the flames, finally shot down seven enemy planes. As the fires burst through the decks of the ships, the soldiers still on board, holding their rifles, jumped over the side. Kept afloat by the lifejackets with which they had been equipped, some managed with difficulty to get into boats while others swam towards the shore.

For these men it was a grim introduction to war.

Groups of enemy fighter planes attacked our launches and poured a hail of bullets into them as they drifted on the surface of the sea, but nevertheless by degrees most of our men got ashore and formed a line on the beach. There, as daylight came, it became impossible to move under the heavy enemy fire at point-blank range. Officers and men instinctively dug with their hands into the sand and hid their heads in the hollows. Then they burrowed until their shoulders, and eventually their whole bodies, were under cover.

Their positions were so close to the enemy that they could throw hand-grenades into the loopholes in the pillboxes. All the time they were using their steel helmets to dig their way further forward, with their swords dragging on the sand beside them. Eventually they reached the wire entanglements. Those with wire-cutters got to work, but they had scarcely commenced when there was a thunderous report and clouds of dust flew up completely obscuring the view for a time. The attackers had reached the British mined zone. Moving over corpses the wire-cutters kept at their work. Behind them followed a few men, piling up the sand ahead of them with their steel helmets and creeping forward like moles. The enemy soldiers manning the pillboxes fought desperately. Suddenly one of our men covered a loophole with his body and a group of the moles sprang to their feet in a spurt of sand and rushed into the enemy's fortified position. Hand-grenades flew and bayonets flashed, and amid the sound of warcries and calls of distress, in a cloud of black smoke, the enemy's front line was captured.

Masanobu Tsuji, *Singapore: The Japanese Version,* translated by M. E. Lake and edited by H. V. Howe, Ure Smith, Sydney, 1960, pp. 93–5.

94
A Successful Ambush of the Japanese

THOMAS LEWIS

The rapidity of the Japanese advance down the Malayan Peninsula exposed its lines of communication to attacks such as that here described. This episode also illustrates the (wasted) potential of the FMS Volunteers whose local knowledge enabled them, in raids such as this, to move unobserved, using speed and surprise (as the communist attackers—see Passage 99 below—used it during the Emergency) to offset their inability to stand and fight, and yet permitted them to cause disruption on a scale out of all proportion to their numbers.

T. P. M. Lewis (elder brother of the author of Passage 54 above) was a member of the Education Service, serving in Perak at the time of the Japanese invasion. He here describes a raid behind the enemy lines by a platoon of the FMS Volunteer Force (see Passage 92 above). This is a preliminary to his account of his three and a half years in internment. He refers to a book by Rose, an Australian officer, who was also involved.

THE two motor boats carrying No. 1 Platoon made their way without incident up the Trong River at a speed of 5 knots and reached the entrance to the creek leading eastwards to the Trong jetty in good time. But to put themselves 2 miles nearer to the Taiping Pass, which was still their target, they decided not to enter this creek but to go further north and enter another tributary of the Trong River called the Langat River. It was just after they had entered this tributary that their luck deserted them. One of the motor boats broke down and refused to start and this so delayed matters that they had to give up any plans for reaching the Taiping Pass that evening, particularly as it took the platoon 3 hours to get through half a mile of mangrove swamp.... After one or two false starts, because their maps, 1915 versions, did not show all the jungle paths, they reached Temerloh village where they decided to spend the night preparatory to launching their ambush early next morning. Major Rose himself went off on a private 'recce' of the ambush area before it got dark, reached the road and from a concealed vantage point surveyed the traffic.

First he watched 6 or 7 lorryloads of Jap troops pass by, then by way of contrast, a signal maintenance party of 6 under an N.C.O. in another lorry who proceeded to test the telephone wires only a few yards away. After that a party of 60 cyclist troops wearing their usual varied assortment of uniforms and headgear cycled past. The Major then returned to join the others and sleep in the coolie lines before making an early start next morning. But sleep was to be only fitful because of the mosquitoes and the party got up at 3 a.m. and were on their way by 4 a.m. By 7.30 a.m. they had reached the edge of the road and the 3 sections of the platoon had been positioned along a straight piece of road on the west side....

The platoon kept a watchful eye on what passed on the road. The first vehicles seen were a fleet of touring cars, transporting white suited locals accompanied by the odd Jap in green uniform. These villagers may have been, as Rose concludes, 'an assortment of Quislings' [collaborators], but all the evidence at this stage pointed to the opposite conclusion that they were just frightened ordinary villagers being forced to comply with whatever orders the invaders gave. Then came two enormous Jap ambulances looking very much like huge furniture vans. Finally when Rose had just flattened himself against a tree at the approach of more traffic, he suddenly sighted the approach of a touring car flying a large blue pennant which indicated that the occupant was a high ranking Jap officer. 'A brigadier, we must get him!,' he exclaimed, and within very few seconds he had taken aim, pulled the trigger and shot the driver dead. The Brigadier's car crashed into a ditch and was quickly followed by a whole fleet of cars which must have been following closely behind. Tommy gun and rifle fire as well as our grenades must have taken a heavy toll of the Japs in the 3 staff cars and the 5 following lorries which they counted. But Major Rose says that he was disappointed that there had not been more control of our fire power in this action and that 2 or 3 more ambushes were not organised in the same area before the Japs had had time to mount a counter offensive. However the Australian command decided otherwise.

T. P. M. Lewis, *Changi: the Lost Years—a Malayan diary*, Malaysian Historical Society, Kuala Lumpur, 1984, pp. 32–3.

95
The Black Market in Japanese-occupied Malaya

CHIN KEE ONN

The Japanese who occupied and governed Malaya and Singapore for three and a half years had considerable difficulties to contend with—especially as the tide of war in the Pacific turned against them. But their corporatist economic arrangements simply opened the doors for a black market to flourish in a country where goods were in ever decreasing supply. Chin Kee Onn's book, published in the immediate aftermath, soon became a classic account of Japanese rule—all the more so because of its derisive humour.

THE Black Market was nurtured by several factors; firstly, shortage of commodities accentuated by the keen competition among Siamese buyers who came to Malaya; secondly, the corruption of government departments; thirdly, the commandeering of transport facilities; fourthly, the functioning of Kumiais [monopolies]; and lastly, the effectiveness of the Allied blockade of Malaya.

Early in 1942, adventurous Chinese traders from Siam, came over to Malaya in junks and lorries, laden with rice, brown sugar, onions, garlic, dried chillies, and condiments for the Malayan market. In return they collected all sorts of foreign foodstuffs, foreign piece goods, foreign industrial implements and raw material. Keen buying by Siamese traders made prices soar. Local speculators pulled the wires and played on supply and demand.

Government departments in every State, realising the exodus of essential commodities, at once set up stringent prohibition export-lists. But all those regulations were blunted by bribery and corruption. Export-permits and Transit-permits could be bought. Military officials, government officials, the Police, the Customs, the detectives, and Transport officials 'closed one eye' on presentation of 'tea-money' or 'protection-money'. All these payments the merchants added on to 'costs'. Every other excuse, such as increased transport or labour charges, was added on to 'costs'. The basic argument of traders who wanted to justify themselves were: 'Look at our costs. look at our risks. Look at our outlay of capital!' By that they meant that the increase of costs had made it utterly impossible to conduct business in the open market at controlled prices.

Traders from Malaya who went to Siam to purchase foodstuffs, were not only victims of the exchange disadvantage, but also of all sorts of corruption on the Siamese side. Consequently, when they returned with rice, sugar, salt, cigarettes, and other goods, they were compelled to have resort to the Black Market in order to make their profits. They also talked of 'costs, risks and poor return'.

Rail transport facilities were considerably curtailed on grounds of 'military requirements first'. Shortage of rail transport coupled with shortage of lorry-transport (lorries and cars having been requisitioned by government and certain Kaishas), meant that ordinary business people who had no official pull could do little in trade. There was therefore no worthy competition against the powerful combines who had 'the necessary influence', and they were at liberty to 'shout prices'. When it comes to foodstuffs, you have either to turn to the Black Market or starve....

There was even a black market within the gaols, the internment, and the POW camps. In gaols, prisoners who had exhausted their money, bartered coats, trousers, shorts, shirts, singlets, and sarongs for food (and that meant extra congee, sweet potatoes, and tapioca). Many gaolers, guards and cell-attendants accepted from prisoners, I.O.Us or letters containing instructions to relatives amounting to 'letters of credit' for food, cigarettes, and other small items supplied them from time to time.

Chin Kee Onn, *Malaya Upside Down*, Jitts & Co, Singapore, 1946, pp. 35–7.

96
A Girls School during the Japanese Occupation

TILLY DE SILVA

From humble beginnings (in 1913) the Pudu English School for girls in Kuala Lumpur rose to become—in the inter-war years—one of the best of its kind in Malaya, notable in particular for its pioneer work in the teaching of natural science at higher levels. As it grew the school was able to finance the construction of some excellent buildings, which attracted the attention of the Japanese as soon as they arrived in Kuala Lumpur. Hence, the school, as a school, had to find alternative accommodation.

Josephine Foss, the headmistress who had done so much for the school, went into internment and, returning to Malaya at the age of sixty after the war, was obliged to find other outlets for her abundant energy and talent (see Passage 100 below). Not the least of her achievements was the quality of the teaching staff, among whom Mrs de Silva, a member of the staff since 1925, was outstanding—as this passage shows. Many years later, in 1988, a history of the school and biography of Miss Foss was written, to which Mrs de Silva, still flourishing, contributed an account of the Japanese occupation period, dictated to a former headmistress of the school at a later time.

THE buildings of the Pudu school were of such quality that the Japanese selected them for use as a military camp. Thus, throughout the occupation period, the school was in exile.... For the next three years the fortunes of the School were in the hands of Mrs de Silva, who had been a member of the staff since 1925.

Mrs de Silva ... as senior staff member with six others, began the registration of students at Tsun Jin Chinese school, a private Chinese school on Loke Yew Road, about a mile from Pudu English School. Students were sorted out by age for different classes. Time-tables were drawn up together with tentative courses of study. All this, together with meetings, took place at the Chinese school.... In May 1942, or thereabouts, the Pudu English School re-opened as an afternoon school in Bukit Bintang Girls School buildings. Mrs de Silva was appointed headmistress, with six teachers, one clerk and one peon (office-boy). Cleaners and gardeners were under Bukit Bintang Girls School employment. A military officer was in over-all charge—Komatsu San was our man—tall, lean and very fierce-looking. Time-tables were altered to include Nippon-Go [Japanese], Japanese songs, music drill (P T), Arithmetic in Japanese; otherwise English was the language of instruction. About seventy students enrolled.

All teachers had to attend Japanese Language classes. A military officer, English-speaking and American-educated, was the instructor. Hashimoto San was a kindly, understanding gentleman, who followed up his classes with offers of help in the various schools. The Pudu school took all its problems to him, even to the point of settling our differences with Komatsu San, who soon softened his stern manner towards students and teachers.

Salaries were fixed by the Education Department. The Director of Education was a high-ranking military officer, very conscious of his position. Rations of rice and sugar at controlled prices were given to all staff members. These had to be collected in bulk and shared out equally in the school. This was a difficult, but a better, method of distribution than the coupon for collection, issued for buying fish and meat once a week at the market, where rubbish in little made-up bundles was offered. The black market sold the best food but this was expensive—an egg cost 35 cents.

In time, as confidence built up, enrolments increased. The children seemed to be enjoying school again. Japanese music in song and dance thrived. Colourful books, sent in by the Propaganda Department, were keenly accepted. The Japanese have a way with children. Ours now looked healthier and happier.

School began each day at 1 p.m. with assembly and a salute to the flag. Teachers and pupils had to learn that bow. Then followed Taiso for all—exercises in drill to a count by the master. Each week a new set of exercises was added, until a 10 minute session could be gone through without a hitch. Then a Banzai shout was given before filing out to classes. A break for tuck was at 3 p.m.; then from 3.20 to 5.30 class resumed, or children had field games from 4.30 to

5.30 p.m. At the end of the year, a Christmas party, with a short programme of Japanese songs and dances, was held, and each child received a little gift from the tree. Christmas carols were sung in English with a tableau of the Nativity on the Hall stage. The party ended with singing Auld Lang Syne in Japanese, as taught by the master....

Then, in 1943, at the end of the first term, at the request of Komatsu San that the school repeat its little end-of-year entertainment for the Japanese, the staff agreed—but on condition that this was held in a public hall for the benefit of everyone. With the permission of the Education Department, the Pavilion Cinema hall was booked and a concert held for two days—a dress rehearsal for the public and the next night for the Japanese. This was the start of other organisations for giving public performances, which did much to bolster the otherwise dull life of the people. Cinemas were expensive, and programmes heavy with propaganda. Night life, like cabarets, were for the few. People were afraid to own radios.

In 1944, the Education Department amalgamated the Bukit Bintang Girls School and the Pudu English School into one school, owing to the fall in enrolment. Mrs de Silva was appointed headmistress, and Komatsu San was the over-all head of Language and Culture. Overnight the Pudu school became the Bukit Bintang Girls School, operating as one in the morning from 8 a.m. to 1.30 p.m. Bukit Bintang Girls School kept its identity as a more efficient school in the Japanese language, both spoken and written. A teacher with a bent for music—piano, guitar, accordion—was transferred to the school. She was an asset, as she was single and a willing worker. Taiso became more interesting, as the teachers and children swung to the rhythm of the music. Komatsu San was pleased.

Soon things in the country began to go wrong. One day a makeshift wooden overhead contraption was erected across Bukit Bintang Road, at the Pudu junction. Five human heads were atop the 'bridge'. Many children turned back home at the gruesome sight. Attendance at school suffered. Enrolments dropped further. Soon after this, the school had to give up three groundfloor classrooms. These were taken over by the Department of Food Control as offices. Later the headmistress, Mrs de Silva, sent in her resignation from the service—against the wishes of the Japanese teacher in charge. The reason for her resignation was that she could not present herself for a second oral examination in Japanese, she having failed the first. The result of this was instant dismissal by the Director of Education, who claimed that her action was an insult to the Administration. One could not tell the Japanese one was leaving. You had to wait to be told to go. To degrade her further he sent her with a note to the Food Controller, asking that she be made a clerk in his office. Thus, she found herself back in the Bukit Bintang Girls School premises the same day, but this time in an office where she was very happy—she was a woman in a room of understanding local men.

Some six months later, she ended up with a sack of rice and a big bag of sugar, given to each of the staff of the Food Control Department by the Japanese, who were emptying the godowns before capitulation.

'Mrs de Silva's Story', in J. M. Gullick, *Josephine Foss and the Pudu English School: A Pursuit of Excellence*, Pelanduk Publications for PESOGA, Petaling Jaya, 1988, pp. 131–4.

97
Radio Josephine

SYBIL KATHIGASU

At the start of the Japanese Occupation Dr and Mrs Kathigasu moved from Ipoh to the outlying township of Papan, where the doctor, in addition to his normal medical practice, secretly attended sick and wounded men from the anti-Japanese resistance.

Mrs Kathigasu, daughter of an Irish planter and an Indian mother, showed much courage and intelligence in her use of a secret radio to receive and disseminate allied broadcast war news. But it did not save her from detection and torture by the Japanese. At the end of the war, barely alive, she was sent to London for medical treatment, but she died there. While the memory of Japanese atrocities was fresh, Mrs Kathigasu, honoured with an OBE, was a national Malayan heroine. After her death her diary was edited for publication. Barh was an electrical engineer who lodged with the Kathigasus; William (Pillai) was their adopted son.

IN the end it was agreed that we should obtain a set, install it in the house, and spread the news among our friends. We should have to work with the utmost discretion, and every member of the household would have a part to play.

I started to make covert enquiries among people I could trust, and soon learned that Mr Wong, a Chinese neighbour of ours, could let us have a set. I saw Mr Wong and agreed to take over his set: he refused payment, and I did not enquire where the set came from. It was a handsome six-valve G.E.C. model. I promised him that we would pass on the news to him, but he was to use the utmost discretion in spreading it further. So the set arrived—brought to the back door late one night. It was forthwith christened 'Josephine', and from then on we always referred, for the sake of security, to 'Josephine'....

As I have said, Josephine was disconnected when not in use, her valves removed, and she herself concealed among packing cases in a corner of the room. One evening Barh was away in Ipoh, and I attempted to connect up Josephine when the time came to tune in to the B.B.C. Alas! the connections I made were the wrong ones. All that issued from Josephine instead of her familiar whisper was a smell of burning and a thin trickle of smoke. On his return Barh examined her and pronounced the patient beyond hope of recovery; he performed the post-mortem and dismembered her so that Dominic and I could give her a decent burial in the family vault.

We did not waste time in mourning her loss, but set about finding a successor without delay. Mr Wong, to whom we first applied, was very sorry, he had no other set, but George Matthews might have one. I saw George after mass on Sunday and tackled him. He knew of several sets which were available, and were never used as there was a strong Japanese garrison in the town, and houses were frequently searched. I decided to take a very small set which belonged to Father Cordiero and was concealed in the ceiling of a schoolroom. It was small enough to be easily hidden and Mr Wong brought it on his bicycle to Papan when he returned from work, hidden in a basket below a layer of vegetables.

However, we were due for further disappointment. Having no short-wave range the midget could not do Josephine's work, and was useless to us. I took it back to Batu Gajah, in a basket of bananas for Father Cordiero. Then I arranged with George to pick up his own six-valve G.E.C. model, which was hidden in an outhouse in Father Cordiero's garden beside the church. This I took back to Papan in the car.

Josephine II was soon working regularly, though before long she needed a new valve, which William procured in Kuala Lumpur, a hundred and fifty miles away, where we hoped the purchase could not be traced. We then decided that she ought to have a better place of concealment when not in use, as the Japs were frequently visiting houses in Papan and searching them. So far they had left No. 74 alone, but we could never tell when our turn might come.

We chose the cupboard under the stairs for Josephine's new hiding place; it had the advantage of brick walls on three sides, which reduced the risk of any sounds from the set being heard in other parts of the house. However, this dark cupboard was just the sort of place the Japs made for when searching a house, and we decided that for safety a hole should be dug below the concrete floor. Most of the work was done by Barh. With a chisel and hammer he laboriously chipped out a square section of the concrete, while Dawn drowned the noise by singing loudly to the accompaniment of tin cans beaten with sticks. The hole when dug was lined with cement, and a cement slab made to cover it.

When not in use Josephine was kept in the hole under the floor with some old sacks thrown, as if carelessly, in the corner of the cupboard to conceal the slab. At the time of a broadcast Barh would lift her out and connect her to various wires which he had cunningly fixed so that they entered the cupboard through a hole drilled in the stairs above. The aerial was led outside the house, and ran up the wall concealed behind the vertical galvanised iron drainpipe.

All went well for a time, but Josephine's troubles were not at an end. One night there was exceptionally heavy rain, and in the morning we found Josephine standing in several inches of water. Of course she refused to function when we tried her out. Barh did what he could, but without success, and we decided to call in Raja, an Ipoh radio mechanic. Raja had been a close friend and we felt we could trust him. He was working for the Japs now, and it would probably have cost him his head if his employers discovered the purpose of his visit to Papan. But he did not hesitate, and William drove him from Ipoh in the car. After examining her, Raja told us that Josephine needed a new transformer, which he was unfortunately unable to get us. On leaving he took with him a bottle of medicine labelled for his wife to provide an excuse for his journey.

We succeeded in replacing the transformer, with the help of our good friends the Catholic Brothers in Ipoh, but Josephine remained a weakly specimen and was often in need of treatment. We decided that she should have a replacement on the spot, so that we might never go without hearing the news; making enquiries through George Matthews I learned of a Philco set which we could have. This was brought to Batu Gajah in a wooden crate, labelled as medicines.

But first, before we could receive Josephine III, we had to prepare a place for her reception. A similar hole to the first was dug in the room at the back of the house where we treated the sick guerillas; this, like the other, was covered with a cement slab, and over the slab we placed a wooden couch, shaped like a great box, on which the patients lay for examination and attention. Both holes were made waterproof. In the second hole we placed Josephine II, who was now a reserve, together with spare valves in a tin. Josephine III was removed from her crate, stripped of her wooden casing and loud-speaker and lowered into the hole in the cupboard under the stairs. The polished casing we burned, and buried the loud-speaker with the other parts in the garden. Josephine III needed a visit from Raja before she was quite fit; thereafter she functioned perfectly and never failed to bring us news of the outer world.

Sybil Kathigasu, *No Dram of Mercy*, Neville Spearman, London, 1954, pp. 49–50 and 81–4.

98
The British Return to Malaya—September 1945

JOHN GULLICK

Until the atomic bomb brought the war with Japan to an abrupt end in August 1945, British war strategy was based on the assumption that the recovery of Malaya would entail a hard-fought campaign, such as had just ended in Burma. The standard arrangement in all theatres of war was that for a few months, at least, the military commander would govern the country across which his lines of supply and communication (with the operational front) would run. For this purpose a British Military Administration for Malaya (and another for Singapore) was formed in India as part of the army's supporting troops. Some BMA personnel had been in Malaya before the war, but the majority had not. All were troops under military command.

When the Japanese surrendered it was too late to modify these elaborate plans. Troops, including BMA personnel, were landed at Morib on the Selangor coast (and some other points), whence they were deployed under the original plan. One BMA detachment was assigned to each State or Settlement of the pre-war constitutional structure. The Negri Sembilan detachment moved from Morib down the coast to Port Dickson, and so on to Seremban. This passage refers to three officers of the advance party who reached Seremban, where they were later joined by about ten more. James Calder, a pre-war Malayan Civil Service officer, had served in Seremban and was in command.

MEMORY is a selective thing. Some impressions last and others fade. Our run in a jeep from Morib beach to Seremban by the coast road was my first trip through the Malayan countryside. I can remember that it was a fine sunny afternoon and (I believe) Thursday 10th September 1945. As we came down into Negri Sembilan we passed through the endless miles of what—much later in my days with Guthrie—I knew so well as the vast Tanah Merah rubber estate. At the Lukut junction we turned inland and so came to Seremban in the late afternoon, as the government offices were closing. It all seemed quiet and peaceful enough.

The first job of a British Military Administration on taking control of a new area is to post the proclamations. These are a set of legislative declarations by which the commander-in-chief (Mountbatten) announces that his writ has begun to run and what (on essential points) it is. As the pre-war treaties between Britain and the Malay Rulers had not then been rescinded, some importance attached to this formal process.

We had come equipped with several copies of the ten or a dozen proclamations which we had to promulgate. We also had a tin of gum arabic crystals with which we could make gum for sticking them up—if only we had some hot water. By this time we were half way down the hill from the Residency to the Railway Station. Calder had an idea. We went on down to the railway station, where we found the Indian stationmaster in a state of high satisfaction. Sanders (the General Manager designate of the Railway) had just been through on a special train and had confirmed the stationmaster in that office under the new regime. All he needed now was a few trains—there were no regular services and timetables were irrelevant.

So we borrowed his pot of glue and a clerk or two and pasted up our proclamations on the timetable boards outside the station. The stationmaster's satisfaction with his lot was now beyond expression in words—and we had put Mountbatten and ourselves on a proper legal basis in Seremban.

The situation in Negri Sembilan at this time was not easy to get a grip of. The Japanese were still there—the Japanese military governor was packing up his bags at the Residency (and complaining bitterly at being restricted to two suitcases—one more, he was told, than his lot had allowed our people to take into captivity in 1942). There were Japanese troops in various places uncertain to whom to hand over their arms—and anxious to avoid a misunderstanding if they kept them meanwhile.

The 23rd Indian Division had landed at Port Dickson a day or two before and its patrols had gone inland as far as Kuala Pilah. The third element in the situation was the Chinese communist Malayan Peoples Anti Japanese Army (MPAJA) who had just emerged from the jungle and begun to move into the towns. With them were the handful of British officers of Force 136, whose task it was to maintain liaison between the MPAJA and the British troops and generally to persuade the MPAJA not to arrogate to themselves an authority which the British authorities were not prepared to concede. It was—and for some time it continued to be—an uncertain and uneasy situation which could easily get out of hand. There was no established and recognised government authority and no one knew who would come out on top....

The Japanese bequeathed to the BMA a number of ills. The entire economy was in a state of near collapse. 'Banana' currency (so-called because the Japanese dollar notes bore a banana tree emblem) had depreciated to the point of worthlessness; local food production and the rationing of foodstuffs was in chaos; there was a serious risk of outbreak of epidemic disease. We did in fact have a smallpox epidemic in the autumn of 1945 but it was contained by mass vaccination before it got out of hand. The two principal communities—Malay and Chinese—had been set at each other's throats. The whole machinery of administration was grinding to a halt. The Police, misused in support of Japanese security operations, were demoralised and almost useless, with the result that law and order was sliding rapidly down the slope towards anarchy....

There was no obvious or sovereign remedy in these circumstances. The BMA simply mended its fences as the need was seen. For lack of any organised political or military opposition we were in the course of a few weeks more or less accepted.... We also made the most of the force which we had. You may have seen that early silent film (a Buster Keaton or a Harold Lloyd) in which the comedian and friend have to ward off an attack on a fort. So they take one of the long uprights off a ladder and march up and down behind a wall of the fort, with the exposed cross-bars of the ladder projecting above the wall like the rifles of soldiers drilling. The troops in Negri Sembilan had the great advantage of mobility (and of a pre-war road system still in comparatively good order). Each small detachment 'patrolled', i.e. went on round trips of its allotted area in trucks. Mobility made it look a larger stage army than it was. No one wanted a fight and the transient presence of troops had a steadying effect.

J. M. Gullick, 'My Time in Malaya', unpublished memoirs, 1969.

99
Communist Guerrillas under Pressure

CHIN KEE ONN

After giving his devastating picture of Malaya under Japanese rule (Passage 95 above), Chin Kee Onn later used his talents as a writer to describe the situation of the Chinese rank and file, many of them fugitives, rather than ideologues, of the communist-led rising known as 'the Emergency'. But in this case he presented his picture in the form of a novel. The passage below is a vivid description of 'the biter bit', that is, the desperate position of raiders, if surrounded and then battered by the overwhelmingly superior numbers and weapons of the security forces. But Kung Li makes a very valid point when he argues that there was 'tremendous wastage of ammunition' in killing even one of the elusive terrorists. Busu-busu was an aborigine guide.

THEN, when all hope had almost gone, Kung Li and five of his men suddenly turned up one cold, misty morning, moving into camp like ghosts. They were bedraggled, red-eyed, and stinking, and all of them were haggard and exhausted. They went straight to their dormitories—plain wooden huts with thatched roofs. There they put down their weapons and knapsacks, slumped into their sleeping places, and fell asleep. Even the tough and redoubtable Kung Li was knocked out this time. After having made sure that Zam Amat and Ah Fee had returned with their squads, he fell on his bed without removing his dirt-caked uniform, his puttees, or mud-laden boots, and within half a minute he was sound asleep.

As soon as she saw him Loo Moy was transformed from a

state of utter dejection to one of absolute joy. Her first impulse was to run to him and fling her arms about his neck. But such a thing was prohibited, and must never be openly done. In their society the rules of propriety had to be strictly observed, and love when publicly displayed was improper and indecent. Therefore, she had to content herself with watching him sleep.

After a while she removed his boots and puttees and dislodged more than a dozen fat leeches from his calves and toes. Then she removed his tunic and singlet, and seeing several leech-inflicted wounds on his back and ribs, she cleaned them up and smeared sulphur ointment on them. Then she put dry clothes on him and watched over him as he lay asleep. Just to do that filled her with a peculiar contentment; she was overwhelmed with thankfulness that no harm had befallen this man whom she loved better than life itself.

Kung Li and three others did not wake until late next day. So exhausted were they and so soundly did they sleep that nobody ventured to wake them. After they had got up and washed, they asked for food and ate like starved wolves.

That night there was no political class. Instead, after the evening meal, they all went to the large hut that served as a lecture hall and waited eagerly for Kung Li's recital of his recent adventures.

Kung Li knew better than any of his cadres how to put across the big talk and how to hoodwink his listeners—but on this occasion he preferred to let them know more of the truth. He knew by instinct that this was not the time for too much bluff. He and his men had returned late, very late, and they were in a terrible condition. Every one in the Camp could see that, and he realized that this time it would not do to distort facts. So, calmly without any dramatics, he told them what had happened to him and his men.

After a day's march from the scene of the ambush they discovered to their dismay that they were encircled by strong enemy forces. British, Gurkha, and Malay troops were slowly closing in on them, and it was obvious that he and his men were trapped. But Busu-Busu was their rescuer. He knew the jungle so well, and led them so expertly, that they were able to elude their pursuers no fewer than six different times.

Suddenly the enemy changed their tactics, widening and

strengthening their cordon. Then for three successive days wave after wave of planes hammered them mercilessly from the air. Heavy four-engined aircraft flew in relays over their sector of the jungle, dropping loads of bombs, many of which must have been one thousand pounders. But the giant trees of the forest were nature's trenches, and behind the mighty roots of these they crouched for shelter from the whining, rending splinters of the bombs. The dense green jungle was filled with the thunder of aircraft engines, with the whistle of the falling bombs, and the ear-splitting crash of explosions. The earth trembled under them; it seemed to rise and shake itself in fear; but by a miracle they were unharmed.

Then came the strafing attacks. Tremendously swift and powerful fighter planes swooped down one after another to release their rockets or terrific barrages of machine-gun-fire. Fern and foliage were shredded and lacerated, bark and branches were ripped off trees, splinters flew in all directions. The hissing of angry rockets and the blood-curdling whine of bullets were enough to drive men insane with terror. There seemed to be no escape from this rain of death.

It was during one of these strafing attacks that Comrade Ah Kum was killed. Weakened by dysentery, he could not run fast enough to reach shelter.

'But this is the point to remember,' Kung Li insisted, 'the enemy dropped hundreds of tons of bombs and used up tens of thousands of rounds of ammunition, but all they got was just one of us. Think of it, only one man killed for all that tremendous wastage of ammunition. At first we were scared, very scared. But after the second day's strafing we lost our fear of the enemy's planes. We realized that their chances of hitting us from the air were as remote as one's chances of winning a prize in a public lottery.'

He then told them of the four day's of artillery bombardment that followed the aerial attacks. These, of course, were calculated to drive them out into the open. But Kung Li and his men refused to be driven out. They moved from one piece of shelter to another, hiding behind trees or between the huge roots of these forest mammoths. Again they were frightened at first, but after a day or two they got used to the explosion of the shells.

The artillery bombardment was followed by several days of intensive jungle-bashing by infantry. They covered a wide area, but could not beat the jungle or the craftiness of the freedom fighters. The enemy could not discover them. They hid in swamps and in dense undergrowth, at times even in trees. It was Busu-Busu's keen sense of smell and sharp observation of details that saved them time and time again. Twice they almost ran into an enemy patrol, and on four occasions patrols almost stumbled upon them. It was touch and go all the time.

'Ah, if they had had dogs with them we should have been done for,' Kung Li remarked.

Altogether, it had been a nightmarish experience, and Kung Li frankly admitted as much. Each man had reached the last *tahil* of his oats and *ikan-bilis.* Fortunately for them they stumbled upon a patch of wild durian, and this kept them going for days. But, most fortunate of all, the frustrated enemy finally moved away.

Chin Kee Onn, *The Grand Illusion,* Harrap, London, 1961, pp. 17–20.

100
'Not Worth a Bullet'

JOSEPHINE FOSS

Although Josephine Foss (Passage 96 above) could not, after the war, resume her career as headmistress of her beloved Pudu English School, she found other useful things to do by joining the newly established Social Welfare Department, initially to train Malayan welfare staff. Later, on the outbreak of the Emergency, she was drawn into the work of relieving the hardships of dependents of communists awaiting repatriation to China, or caught up in the relocation of Chinese rural communities into 'new villages'. After that, indomitable as ever, she went on to become headmistress of a school in Sarawak, only retiring when she was 70.

THE new villages grew rapidly into towns with cinemas, clinics, shops, coiffeurs, churches. But many of the families living on the outskirts of these villages were in a bad way. As more than one family told me, 'If we do not give them food through the bars at night, they will shoot us, and if we do, the British will put us in prison.'

Another good idea was the Morib camp, not very far from Port Swettenham. When a bandit was caught and agreed to go back to China, he was allowed to take his family with him and the task of those poor welfare officers was to find the families. Then we would house them in the Morib camp until the husband's case was over, so that they could sail for China together. I seemed to be running backwards and forwards to Morib so much, joining up families, perhaps with prisoners I had seen in Pudu or Taiping prison.

I nearly always drove the car myself and I had a faithful Chinese welfare man who always went with me. We will call him Tan. We never had an escort, but I needed him in case the car broke down, or in any other emergency. You see the bandits were getting short of ammunition and I was 'Not worth a bullet'. Once when Tan and I were taking a lorryload of women down to Morib, Tan and I in the car, the lorry broke down and the women began singing 'The Red Flag' at the top of their voices.

I dared not leave them, as they might persuade the lorry drivers to turn back (or perhaps bribe them, for many of them were well off). The car bringing the Bishop of Singapore, Bishop Wilson, from a Confirmation in that area, happened to pass and his driver, a Malay and a good friend of mine, pulled up and asked what he could do. They kindly offered to drop Tan at the nearest police station and I was left with the lorry and 'The Red Flag'. Help came very soon, and also another lorry, and my ladies were soon housed in the camp at Morib. Many had bicycles, sewing machines and good clothes.

Another task we had to do was to feed the bandits on the train, when it stopped in Kuala Lumpur. Quite a number of our welfare workers helped and the Chinese contractor who supplied the rice always did it well. They were handcuffed together, but I always got permission, while they were eating in groups, to have their handcuffs taken off. It is very hard to use chopsticks if you are handcuffed. They never let us down—'Honour among thieves'. One day, as we chatted with them, a man asked me if his wife was in Morib, and I said how could I know, as I did not know his name. He replied, 'But all the Communists know you!' What a reputation! She was safely in Morib and soon away to China. Actually, nearly all these were—what one would call—the lighter cases, young men picked up in a round-up or given away by Sakais etc.

And that leads me to round-ups. This was the most horrible part of my job. Sometimes it was comparatively easy, as in the following case. There had been a lot of trouble at Kajang, a small town near Kuala Lumpur. Some Scots troops had been sent in to round-up the village, where most of the Chinese lived and where they had done much harm to the rubber estates nearby. The Communists always found out beforehand, and when the troops arrived there were only old men and women and children there.

As there was some trouble and Tan was not available, Dr Rawson, the head Welfare Officer, asked me to go along as it was not far. I got there to find a crowd of screaming, yelling Chinese women and the troops not knowing how to quiet them. I spoke to one woman on the edge of the crowd and asked what was the matter. She said, 'These horrible tall men won't let us feed our babies or go to the loo.' I spoke to a harassed young Lieutenant and told him to take all his men out of sight, just leave one I could call in emergency. But there was no trouble at all. As soon as the men went away, coats opened and babies quickly breast fed. Others wandered off to relieve themselves.

I asked why they were afraid of the soldiers, and got the answer that they had never seen men so big before. The young officer brought a few married men and they nursed the babies and played with the children, and the government interpreter was able to get a lot of information out of them. All so simple!

When we went out with the Gurkhas it was a much longer process, as they used to round up more jungly villages. Our task as welfare officers was not with the villagers but to run a canteen for the troops themselves. None of these tasks was easy because all round-ups were necessarily secret, and we had to collect supplies, stoves etc. to be ready to start at the word 'Go', and often I did not know where we were going. I was simply told 'At 5 am tomorrow, there will be canteen work, take your car and Tan; supplies will be sent, with other welfare workers, in a lorry.' Tan always seemed to know, before I got my secret instructions the next morning.

The Gurkhas were wonderful to work with, so kind, grateful and sympathetic to us. The only thing I hated was when, with glee, they brought in a dead bandit, often so pathetically young....

[In another case] it had been decided to burn down a village where there had been constant attacks and many planters had been killed or wounded. We, the Welfare Department, had to collect the inhabitants, and as many of their goods and chattels as were necessary, and put them into three lorries and send them to a detention centre, where they would have to stay till we collected their relations. Then to Morib and China.

It was a small village street with lots of tiny shops, and only women, children and old men were there.... I went on a tour and found a small Methodist chapel, where the bandits must have hidden—remains of food etc. about, and Bibles, hymnbooks, torn and thrown anywhere. I took all the religious books that were salvageable and put them into the car for return to the Methodists later. I know I had a lot of trouble about a Chinese medicine shop—the Malay officer said that those drugs are poison, but knowing how the Chinese love their own cures I managed to persuade the officer to allow the woman to take some with her. The point I could not deal with was the money she had—quite a lot—she must have been selling drugs to the bandits, but I hoped they allowed her to keep some of them....

A little girl had two pet rabbits and was crying bitterly because she could not take them. 'Certainly not,' said the officer.... Hardly had we started on our way, when the troops began to burn the village and the poor things saw all they had lived with go up in flames.

Josephine Foss, 'Not Worth a Bullet', unpublished autobiography, *c.*1960.